W.A.T.E.R.

West African Technology, Education and Reciprocity

By Bradley Striebig

Contributors: Jessica Oddo, Nadia Warren, Sushil Shenoy, Alex Maxwell, Gilbert Nalelia, Jessica Flanery, Laura Cain, Kayla Latimer, Pia Longinotti, Annie Luu, Justin Meeks, Maleena Scarsella, Susan Norwood, Mary Jeannot

W.A.T.E.R.

First Edition

ISBN 978-0-6151-9491-2

To order visit www.lulu.com

For more information on the WATER program visit: web.mac.com/water_dr

Published by
Bradley A. Striebig
Spokane WA
bstriebig@gmail.com

Acknowledgements

The author would like to thank all those that contributed time, money and effort to the WATER program. The author work like to acknowledge the efforts of the other faculty members, Susan Norwood and Mary Jeannot for making this program possible. Other contributors included Mark Alfino, Burt Cohen, Therese Covert, Jeff Hazen, Gary Weber and the staff at Gonzaga University who support the related programs. The officers of EWB-USA and EWB-Gonzaga also played critical roles in the foundation and execution of the WATER program. The author is forever indebted to his parents for providing the foundations for the belief in this work. Finally, the author must thank his wife Abigail and children Preston and Zachary for providing the courage and inspiration for the WATER program.

Thanks to the Faculty and Staff at Gonzaga that have approved of and assisted in the development of the WATER program including:

Father Robert Spitzer, President of Gonzaga University

Thayne McCulloh, Ph.D., AVP

Dennis Horn, Dean of the School of Engineering and Applied Sciences

Mary McFarland, Dean of Professional Studies

Professor Noel Bormann, Chair of the Civil Engineering Department

Sima Thorpe, Director of the Center for Community Action and Service Learning (CCASL)

Wanda Reynolds, Director of the Study Abroad Office

Raymond Reyes, Associate Mission Vice President

Gina Bowman, Academic Vice President - Administration

Professor Gary Weber, Business

Professor Kay Carnes, Business

Professor Mark Alfino, Philosophy

Professor Terry Gieber, Chair of the Art Department

Professor Joanne Smeija, Chair of Chemistry Department

Assistant Professor Dan Garrity, Broadcasting

Associate Professor Phil Appel, Mechanical Engineering

Assistant Professor Matthew McPherson, Business

Professor Anwar Khattak, Civil Engineering

Professor Paul Nowak, Civil Engineering

Associate Professor Sara Ganzerlli, Civil Engineering

John Dacquisto, Head of Center for Engineering Design

Professor Bob Stiger, Mechanical Engineering

Therese Covert, School of Engineering and Applied Sciences

Katuska Kohut, Study Abroad

Donna Ryan, Study Abroad

Joann Waite, Director of Sponsored Research and Programs

Denny McMonigle, Office of Sponsored Research and Programs

Dianne Farrell, Office of Sponsored Research and Programs

Carol Bonino, University Relations

Pete Tomey, Public Relations

Dale Goodwyn, Public Relations

Carie Schwede, Assistant Dean of Admissions

The WATER program is also indebted to Ron Rivera of Potters For Peace and Burt Cohen of Potters Without Borders for their assistance and training that helped us learn how to make the ceramic water filters. We also appreciate the support of Rotary International for the program, in particular the members of Spokane's Rotary Club 21.

This program is grateful for the grant support provided by the United States Environmental Protection Agency's People, Prosperity and Planet (P3) program. The faculty would like to express their appreciation to Julie Zimmerman and Cynthia Nolt-Helms for their dedication to the P3 program.

Thanks also to all the many individuals not mentioned by name who contributed to the W.A.T.E.R. program.

The information and opinions expressed within the book are solely those of the author and contributors and do not necessarily represent the opinions of program sponsors.

Participants

2007 WATER Participants

Elaine Brown, Nursing

Laura Cain, Nursing

Renae Dougal, Nursing

Jessica Flanery, Broadcasting

 Crystal Humphreys, Education

Paul Krupski, Mechanical Engineering

Kayla Latimer, Civil Engineering

Pia Longinotti, Education

Annie Luu, MBA

Alex Maxwell, Civil Engineering

Justin Meeks, Civil Engineering

Jessica Oddo, Nursing

Christopher Pavese, General Eng.

Maleena Scarsella, Civil Engineering

Sushil Shenoy, Civil Engineering

Sarah Vacanti, Nursing

Nadia Warren, MA/TESL

Academic Year Senior Design Participants

2007-2008

Christa Fagnant, Civil Engineering

John Gilliland, Mechanical Engineering

Dustin Hannafious, Civil Engineering

Kayla Latimer, Civil Engineering

Justin Meeks, Civil Engineering
Ashley Parrish, Mechanical Engineering
Kim Remick, Civil Engineering

2006-2007

Natalya Avdyev, Civil Engineering
Andy Elder, Mechanical Engineering
Rachel Kane, Civil Engineering
Annie Luu, Civil Engineering/Business
Luis Martinez, Mechanical Engineering
Chris Oodle, Civil Engineering
Chris Pavese, General Engineering
Nick Ranvenna, Civil Engineering
Keith Rudie, Civil Engineering
Sushil Shenoy, Civil Engineering
Joe Stippel, Civil Engineering

2005-2006

Sam Atwood, Civil Engineering
Jake Dial, Civil Engineering
Blake Johnson, Civil Engineering
Ben Lemkau, Civil Engineering
Kim Remick, Civil Engineering
James Shamrell, Civil Engineering
Phil Spuler, Civil Engineering
Krystal Stanick, Civil Engineering
Amanda Vernon, Chemistry
Julia Young, Civil Engineering

2004-2005

Ian Culbreath, Civil Engineering

Danielle Hardy, Civil Engineering

Katie Hall, Civil Engineering

Tyler Jantzen, Civil Engineering

Brendon Langenhuizen, Civil Engineering

Christa Olsen, Civil Engineering

Mark Raleigh, Civil Engineering

Katherine Rowden, Civil Engineering

Collecting water in Africa

The WATER team at orientation

Preface

"In the age when man has forgotten his origins and is blind to even his most essential needs for survival, water along with other resources has become the victim of his indifference."
– Rachel Carson

The water crisis in sub-Saharan Africa is critical. Lack of clean water is the leading reason that hospital beds in Africa are full, and yet most who are sick from bad water cannot seek medical attention. Addressing the shortage of clean water in developing nations will take investments of money, time, and resources. This is just one of many stories about innovative ways of tackling the water crisis – all of the successful stories involve leadership, community building and personal sacrifice.

Welcome to a new world, one in which Africa leads the way and yet many of its inhabitants still suffer from health issues which were eliminated from the developed world decades ago. What you are about to read is not a thesis by experience researchers in anthropology or African history. This book is about getting to know Africa through naïve eyes, where even the professors take on the role of student.

The W.A.T.E.R. (Water for Africa: Technology, Education and Reciprocity) program at Gonzaga University was born from conversations with students, particularly students in the engineering program. These students came to me with a desire to create positive change in the world, something we all espouse but seldom actually try to do anything about. However, these students were not content to talk about change, as is the personality of many an engineer. So it was that I came to learn about a new organization called Engineers Without Borders, a University rooted program to assist with sustainable development that originated with Dr. Bernard Amedai at the University of Colorado.

Gonzaga students and I agreed to begin a chapter of Engineers Without Borders (EWB) at Gonzaga University. EWB, being a relatively new organization, had rather limited resources and a rather limited list of potential projects in which we could actually get our hands dirty. The only project remotely related to my area of expertise was in a country called Benin. Not a single student or faculty member could say where Benin was without "googling" the name. This being the only option at the time and with students wanting to graduate someday, there was little choice but to accept a potential project on the dark continent of Africa.

In 2004, we were fortunate to receive an inaugural grant for sustainable education from the United States Environmental Protection Agency's P3 (People, Prosperity, and the Planet) program. This grant allowed us to get our first glimpse of Benin, of its beauty and bleakness. It was not until nearly three years later that I was able to return to this magical place and complete a small part of the promise that I made to over 30 students over the course of those three years. Someday, the hard work, talents and education of these students would benefit the people of Benin.

It is only as a result of the hard work and dedication of the students that we were able to listen to their ideas and invent the W.A.T.E.R. program. Everyone that participates in this program owes a tremendous amount to these selfless students who worked on the project based solely upon the inexperienced word of a crazy professor that someday their work would pay off.

It is my hope that this book documents what the people of Benin have taught us.

Contents

Motivation

By Alex Maxwell

It was around 8:00pm as I sat at my desk in an empty office. After a long day of work at my internship and a three hour paddle on the Arkansas River, I found myself pouring over my summer course work for the Benin program. As I sat there thousands of miles away from that small country on the West African coast, I began to study on disease transmission and the water crisis that was impacting everyone around the globe, most specifically the poor. Slowly shaken from my stupor by what I was reading, my heart began to race as my mind wrapped around many shocking facts and figures.

I discovered that 6,000 people around the world die each day due to water-borne diseases. Coupled with this troubling fact, I also learned that out of a village of 1000 people in Africa, over 600 lack access to a latrine, 20 people suffer daily from diarrhea, and getting water for a family of six may take three hours out of a day. As a whole, around 40% of the African population lacks access to clean water and sanitation. Not only that, but malaria, a water-related disease, kills one million people every year, mostly in sub-Saharan Africa, and around 80% of these people are children under the age of five. Along with malaria, I learned that there have been a total of 173,359 cases of the water-borne disease called cholera in Africa, accounting for 94% of the total global cholera cases.

The list of facts went on, and I sat there astonished by how unaware I was of these problems. I was quickly learning how important clean water and sanitation was to the health and growth of developing nations. Having access to clean water either directly or indirectly impacts problems related to poverty, universal primary education, gender equality/women empowerment, child mortality, and diseases like HIV/AIDS. Otherwise left unchecked, lack of clean water traps developing countries in viscous cycles of poor sanitation practices leading to degradation of health and the promotion of disease and poverty.

As I also came to realize, this dreadful cycle cannot be solved simply by sending in more money, supplies, or equipment to places like Benin. Solutions to the water crisis require localized effort and global partnerships with those in need. In making our own localized effort, the Gonzaga University Water for Africa: Technology, Education, and Reciprocity (WATER) program in Benin is joining hands with a larger global community that is committed to addressing the water crisis and making local and sustainable contributions everywhere to solve the problem. This is a global community that has strengthened over the years, thanks in large part to the United Nations Millennium Development Goals – goals that seek unite people everywhere in hopes of achieving a brighter future.

In early September 2000, 189 world leaders came together at the Millennium Summit to discuss the future of the modern world. Out of this meeting came one of the most important declarations in world history, the United Nations Millennium Declaration. Out of the eight chapters of the declaration, eight goals and twenty-one targets, now known as the Millennium Development Goals (MDG), were born in hopes of improving the status of our world by 2015. The main goals and targets of the MDG are as follows:

Goal 1: Eradicate Extreme Poverty and Hunger

- Target 1: Halve, between 1990 and 2015, the proportion of people whose income is less than $1 a day
- Target 2: Halve, between 1990 and 2015, the proportion of people who suffer from hunger

Goal 2: Achieve Universal Primary Education

- Target 3: Ensure that, by 2015, children everywhere, boys and girls alike, will be able to complete a full course of primary schooling

Goal 3: Promote Gender Equality and Empower Women

- Target 4: Eliminate gender disparity in primary and secondary education, preferably by 2005, and in all levels of education no later than 2015

Goal 4: Reduce Child Mortality

- Target 5: Reduce by two-thirds, between 1990 and 2015, the under-five mortality rate

Goal 5: Improve Maternal Health

- Target 6: Reduce by three-quarters, between 1990 and 2015, the maternal mortality ratio

Goal 6: Combat HIV/AIDS, malaria, and Other Diseases

- Target 7: Have halted by 2015 and begun to reverse the spread of HIV/AIDS
- Target 8: Have halted by 2015 and begun to reverse the incidence of malaria and other major diseases

Goal 7: Ensure Environmental Sustainability

- Target 9: Integrate the principles of sustainable development into country policies and programs and reverse the loss of environmental resources
- Target 10: Halve, by 2015, the proportion of people without sustainable access to safe drinking water and basic sanitation
- Target 11: Have achieved by 2020 a significant improvement in the lives of at least 100 million slum dwellers

Goal 8: Develop a Global Partnership for Development

- Target 12: Develop further an open, rule-based, predictable, nondiscriminatory trading and financial system (includes a commitment to good governance, development, and poverty reduction; both nationally and internationally)
- Target 13: Address the special needs of the Least Developed Countries (includes tariff- and quota-free access for Least Developed Countries' exports, enhanced program of debt relief for heavily indebted poor countries [HIPCs] and cancellation of official bilateral debt, and more generous official development assistance for countries committed to poverty reduction)
- Target 14: Address the special needs of landlocked developing countries and small island developing states (through the Program of Action for the Sustainable Development of Small Island Developing States and 22nd General Assembly provisions)
- Target 15: Deal comprehensively with the debt problems of developing countries through national and international measures in order to make debt sustainable in the long term
- Target 16: In cooperation with developing countries, develop and implement strategies for decent and productive work for youth
- Target 17: In cooperation with pharmaceutical companies, provide access to affordable essential drugs in developing countries
- Target 18: In cooperation with the private sector, make available the benefits of new technologies, especially information and communications.

Though the list of goals may appear daunting at first, the global community has been coming together over the past few years to show their commitment toward achieving these goals. It is a commitment that shared by hundreds of countries, companies, interest groups, non-profit organizations, and people working interdependently to give what they can to show the world they care.

In early August, almost seven years after the forming of the MDG's, our small interdependent group of Gonzaga nursing, communication, language, and engineering students and faculty came together in an effort to contribute what we could toward achieving these goals.

Through the implementation of the Filtron, teaching of health education classes, and most importantly, the establishment of a global partnership (Goal 8), our team managed to make a great impact in the lives of those living in Benin. Working side by side with our new friends, we gathered the resources to start the first ever sustainable Filtron production process (Goal 7) in Benin, in addition to teaching some basic sanitation/health practices – finally making clean water a reality to those in need. With clean water, adults, but mainly children, in Benin would no longer suffer from a long list of water-borne diseases (Goal 6). With the decrease in illnesses, goals such as universal primary education (Goal 2), reducing child mortality (Goal 4), and improving maternal health (Goal 5) would be more easily obtained. This new, clean water would also allow for safer food preparation, drinking water, and washing, indirectly aiming toward achieving the eradication of extreme poverty and hunger (Goal 1).

There is hope that our efforts in Benin will reach the lives of thousands in need of help. Though it alone will not solve the problems of the much larger global community, we know we are not the only ones making an effort to help, and our WATER project did and will continue to make the lives of the Beninese people brighter. Most importantly, in traveling to Benin, we showed our friends that they no longer have to be bound in the viscous cycle of poor water quality and sanitation practices that lead to the decline of health and increase of disease and poverty. Instead, we extended them our hands and clasped them tightly, knowing that together we could bring smiles to the faces of many people we may never meet again.

Gathering water is typically left to the children

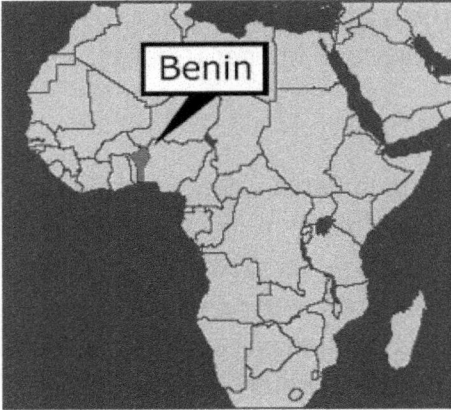

The Songhai Center is located in Porto-Novo, in the southern part of Benin

W.A.T.E.R. for Africa

Water for Africa is an interdisciplinary approach to development attempting to balance Technology, Education and ensuring Reciprocity among project partners. The W.A.T.E.R. (Water for Africa; Technology, Education and Reciprocity) program is designed to meet the needs for a thirsty planet. As part of the WATER program 17 students from eight academic programs and 3 faculty members traveled to Benin in August 2007.

The mission of the WATER program is to provide a service based learning experience to introduce Gonzaga University students and Songhai Center staff and interns to strategies for sustainable development in Sub-Saharan Africa. We believe education is the foundation of improving access to water and reducing poverty and early mortality in children.

The people in rural Benin lack access to clean water. As a result, they live in an environment that fosters many water-borne diseases, such as E. coli and dysentery. These diseases decrease the longevity and quality of life for the Beninese people. The United Nations recognizes that safe drinking water is the first step toward a brighter future for the people of Benin. The goal of this project was to design a manufacturing facility for an affordable, sustainable water filtration system that can be easily distributed throughout Benin.

The program began with a short orientation at Gonzaga University. The purpose of the orientation was to introduce students to one another and the objectives of the course. The course objectives were for students to be able to:

- Describe contemporary health problems in Africa and their contributing factors.
- Describe the relationship between water, sanitation, and causes of morbidity and mortality in Africa.
- Demonstrate communication skills for providing culturally appropriate health education.

Health and Development

The United Nations (UN) has developed the Millennium Development Goals (MGD) program to reduce poverty and improve access to water and sanitation throughout the developing world. The UN has specifically stated a goal to halve the number of people without access to clean water and sanitation by 2015. However, on March 19th, 2006, a statement by the UN noted this goal was in jeopardy in sub-Saharan Africa due to drought, poverty and political factors.

In short, experts in the field of sustainable development believe providing clean water and sanitation for sub-Saharan Africa is one of the world's greatest challenges.

Benin lies in the heart of sub-Saharan Africa and its population lacks access to safe water and sanitation. Centralized water treatment is not a feasible option for community drinking water in Benin because it is extremely expensive to construct and maintain.

In rural Benin, the primary ways in which people obtain clean drinking water are by boiling water or purchasing imported bottled water. Boiling the water requires wood and native vegetation, depleting local resources and emitting smoke into households and the atmosphere. Buying imported water is not a cost-effective, long-term solution for low-income populations. Furthermore, the link between drinking water, sanitation and disease is not clearly understood. As a result, most people in Benin drink water that does not meet the drinking water standards set by the World Health Organization (WHO).

Children in Porto-Novo, Benin

The water in much of Benin is contaminated with bacteria and viruses. These are the major concerns in Benin drinking water because of their impact on human health. Typhoid fever, amoebic dysentery, schistosomiasis and cholera are just a few of the diseases spread by contaminated water. Nearly 17 percent of children born in Benin die before the age of five.

Providing the technology to implement point-source water treatment in the community will likely decrease childhood and maternal mortality rates in Benin. This background along with an assessment trip to Benin by Gonzaga university (GU) in 2004 provided the initiative between GU, Engineers Without Borders USA (EWB-USA), Potters For Peace (PFP) and the Songhai Center to propose a manufacturing facility for drinking water filters. Community focused projects, such as this one, directly address the eight Millennium Development Goals set forth by the UN.

Microbial and chemical contaminants measured in Benin water samples

Contaminant	Units	Concentration in Benin Water	WHO Standard	US EPA Standard
Total Coliforms	MPN/100ml	>1600	0	0
Fecal Coliforms	MPN/100ml	20	0	0
E. Coli	MPN/100ml	NA	0	0
Pathogens	MPN/100ml	>8	0	0
Lead	µg/L Pb	4	10	15
Arsenic	µg/L As	ND	10	10
Nitrates	mg/L NO_3^--N	>30.0	50	10
Phosphate	mg/L PO_4^{3-}	0.19	NA	NA

Appropriate Technology

A ceramic filter called a Filtrón, developed by PFP, was selected for water treatment. The Filtrón is a porous ceramic filter that is designed to fit within a five-gallon plastic pail or clay container. As water passes through the Filtrón, microorganisms and particulates are trapped in the clay. Most bacteria and parasitic organisms are larger than the pore spaces in the filter.

The ceramic filter in a plastic 5 gallon container

Each Filtrón is impregnated with a small amount of colloidal silver, which acts as an anti-microbial to prevent mold from growing on the filter. The silver is painted in minute quantities on the inside and outside of the Filtrón surface after the Filtrón has been fired and cooled. The process to make the filters is described in a later section.

The Filtrón has been cited by the UN in its Appropriate Technology Handbook and used by the Red Cross and Doctors Without Borders. An appropriate technology is a technology that uses the skills and resources that can be found in the local area.

The ceramic filter uses local clay and sawdust as the raw material. A press is needed to form the filters and this press can be made from common materials found in most metal-working shops. A kiln to fire the filters is also needed, the kiln can be the type commonly used in making pottery and bricks. All these resources were identified and available in Benin.

Most other water treatment technologies require more energy (UltraViolet disinfection systems) or chemical additives (Chlorination or other chemical disinfectants). Energy and chemical intensive disinfection systems may provide comparable or even better disinfection, however, the cost and availability of energy and chemical supplies is not sustainable within Benin.

Students at GU have been working to evaluate and improve this ceramic filter technology since 2004. The Filtróns removed more than 99% of the bacteria in laboratory testing at GU. As a part of the orientation weekend, all the WATER team members were taught how to make these simple and effective drinking water filters.

Point of use appropriate water treatment technologies

Technology	Advantage	Disadvantage	Filter Time
Biosand™ Sand Filtration	High removal efficiency for microorganisms	Needs continual use and regular maintenance Cost	1 -2 hours
Filtrón™ Ceramic Filter	High removal efficiency for microorganisms Sized for households Relatively inexpensive	Requires fuel for construction Limited lifetime Requires regular cleaning	2 – 8 hours
SODIS™ Solar Water Disinfection	Highly effective Inexpensive Can reuse a waste product (PET bottles)	Long treatment time (6 to 48 hours) Does not remove other pollutants Requires warm climate and sunlight	12 – 48 hours

Demonstrated reduction in microbial and chemical contaminants with Filtrón ceramic water filters and activated carbon water treatment in the Gonzaga University Laboratory

Contaminant	Units	Filtered water	Average Removal
Fecal Coliforms	MPN/100 ml	< 2 ± 0	>99.92%
Total Coliforms	MPN/100 ml	< 2 ± 0	>99.97%
E. Coli	MPN/100 ml	< 2 ± 0	>99.0%
Pathogens (H_2S producing bacteria)	MPN/100 ml	< 2 ± 0	>99.7%
Streptococci	MPN/100 ml	< 2 ± 0	NA
Amoeba	MPN/100 ml	37,000±115,000	99.5
Lead	µg/L Pb	1 ± 1	73%
Nitrate	mg/L NO_3^--N	11.9 ± 1.5	35.1%
COD	mg/L	25 ± 11	66%

Of course our time in Benin would not be all work. We were able to tour the countryside and experience some of the amazing culture and history of Benin.

Benin has some of the richest and most storied history in West Africa. It is home to the legend of the Amazons and the birthplace of Voodoo. There are two UNESCO cultural sites in Benin marking the impact the Dahomey Empire had upon world history.

The following is a fictional travel diary of a WATER team member experiencing Benin. This account draws upon the actual journals and experiences of several WATER team members:

Departure

Sitting in the airport getting ready to travel to Africa has given me a lot of time to think about what I expect out of this trip. I am not really sure what to think about it all. I am trying to go in without any prejudgment, but it is hard to erase what images and thoughts Hollywood and the news have placed in my mind through movies and articles.

Boy, how am I going to react to this experience? I am not quite sure. I am trying to go into this without expectations because I believe they will be blown away by reality. I do not expect to come back the same person I was when I left, or am now for that matter. I just hope to be able to have an effect on something bigger than myself.

I feel like a child going to school for the first time. I do not know if I feel more nervous or excited. My parents have told me what it will be like, but I am going someplace that I have never been, with a group of people that I do not know, with a language where big, strange sounding words that I will not understand will be used. I know I will learn more than what has been shown to me in picture books, and will experience 'the big world now.'

There's some famous quotation about how we cannot fully see our Earth until we are willing to leave our atmosphere. At first this seems impossible; how can we see things clearly from so far away? But I think this is absolutely true. Until I experience a culture entirely different than my own, I'll never understand my own.

One thing is clear, I have definitely developed a passion for the water crisis. It is something that I have felt completely ignorant about prior to completing our summer course work. I can't wait to see how far our efforts will go towards impacting the lives of these people.

Africa here we come . . .

After eight weeks of coursework and 36 hours of air travel, the WATER team landed in Benin, grabbed the luggage and got on a bus to the Songhai Center.

First Impressions

The Benin airport was absolutely ridiculous! My goodness.
There were so many people and no order. Everyone
wanted the same thing: to find their bags and get out of
the airport. I was glad when we got out of there!

The Benin traffic was almost as shocking as the airport.
Many consider me an "aggressive" driver, but I do not
compare to the drivers in Benin. People on mopeds zoom
past other cars without a worry. Talk about intense!

Traffic in Cotonou, Benin

I didn't really know what to expect of Benin upon arrival. I mean I did have some preconceived ideas from what I had seen in documentaries on the travel channel and Hollywood movies but in retrospect, I really had no idea what I was getting myself into.

Father Nzamujo welcoming us at the Songhai Center

I was so impressed when we first arrived; we were treated like royalty. The Songhai Center where we stayed even prepared a feast for our arrival. We were given private rooms with showers and even air conditioners. Honestly, I didn't expect to even see an air conditioner the whole time that I would be here.

Songhai – A New Vision for Africa

The Songhai Center is located in Porto-Novo, Benin, the colonial capital of the country. The Songhai Center's mission is to promote "the emergence of a new African society with the continual focus on sustainable socio-economic entrepreneurship, the capacity to efficiently harness local resources (natural and human) and the ability to take an effective role in world affairs."

The mission of the Songhai Center is sustainable development through education, technology and sound business practices

Songhai is the vision of Father Nzamujo Godfrey. This vision has turned a former trash dump into an agricultural paradise, with some of the highest product yield per acre in the world. This was accomplished through a creative blend of technology and education about sustainable organic agriculture practices.

The Songhai Center is just amazing. I'm impressed most with their ability to maximize what few resources they have. They treat their own water naturally with water hyacinths. Methane gas produced by waste products is used to cook food. Crops are arranged to use the land most efficiently; they position shady plants under fruit trees, use sunflowers to attract bees to pollinate the crops, and irrigate everything with recycled water. Waste material is used as compost to maintain fertile soil. The director said that the land, which was given to him because it was deemed worthless, now produces better and better crops each season.

For example soybeans are harvested for their oil and replenish the nitrogen in the soil as they grow. The wastes from the soybean plants are used to feed the fish. The fish are used to filter water. The fish are served in the restaurant. Finally the scrapes from the restaurant are composted or fed back to the animals.

There is so little waste here, in fact, that I have yet to find a trash can...

The first taste of fresh pineapple for most of the travelers from Gonzaga

Everything eaten during our stay was grown organically on one of six Songhai Center farms in Benin

While at Songhai, we were served three large meals each day. Everything we ate was produced organically on one of six Songhai farms. The Songhai vision has been extended beyond agricultural production, to encompass all parts of sustainable development.

On a previous visit to Benin, Nzamujo identified the need for an affordable drinking water treatment device for Songhai and the people of Benin. While Songhai is completely committed to sustainable development, they have lacked the capacity to address the tremendous need for clean water in Benin. The WATER program was designed to directly meet the needs identified by Father Nzamujo.

Experience and Education

After getting acquainted with the Songhai Center, we went into the community to collect water samples in order to better understand the necessity for water treatment. The people most affected by waterborne diseases are children under five years of age.

Before I first arrived in Africa, I had these preconceived notions of what the water was really going to be like. I had heard how bad it was and had envisioned brown stagnant water that was pulled from a pond and had little floaters of who-knows-what in it.

Although this seems scary, the truly scary thing is that the water that they are drinking looks very similar to the water that comes out of my tap back at home. The only difference is that the water has a slightly brown tinge to it. But other than that it looks clean, it doesn't smell bad, and to the common eye you really can't tell anything is wrong. Unfortunately, the water is a haven for parasites and bacteria.

I thought that water that kills would look a lot different than water that is safe. Unfortunately, I was wrong.

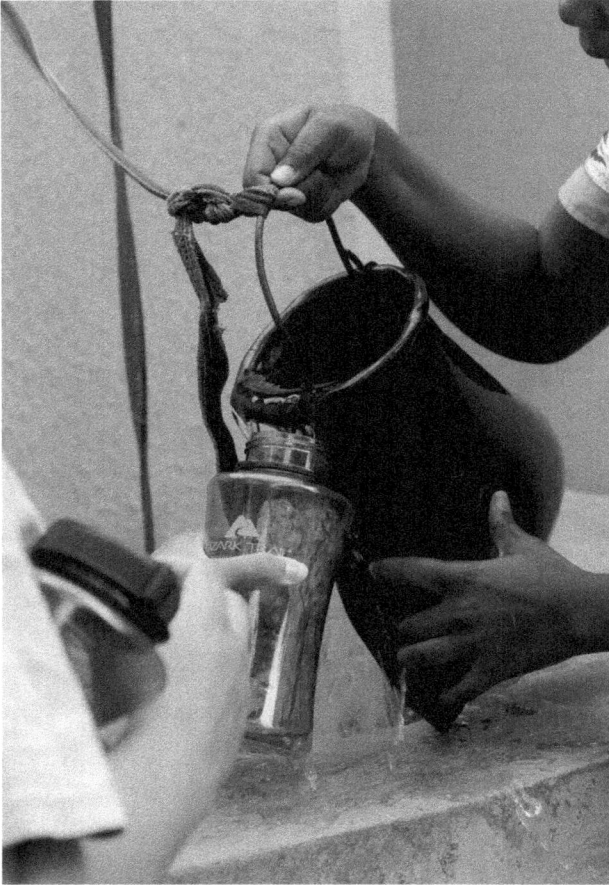

Sampling well water at a home in Porto-Novo

After sampling water at the Songhai Center and homes in the community, the water was tested in a newly established water quality laboratory at the Songhai Center. The equipment was donated through contributions to the water program. We taught Michel and Wilson, the Songhai microbiologist, how to conduct standard water tests.

Unfortunately, nearly all of the water samples failed the basic water quality tests, even the Songhai water. This indicates that the water carries waterborne pathogens and is a source of amoebic dysentery and potentially even typhoid fever or cholera.

Well water that was tested and is contaminated.

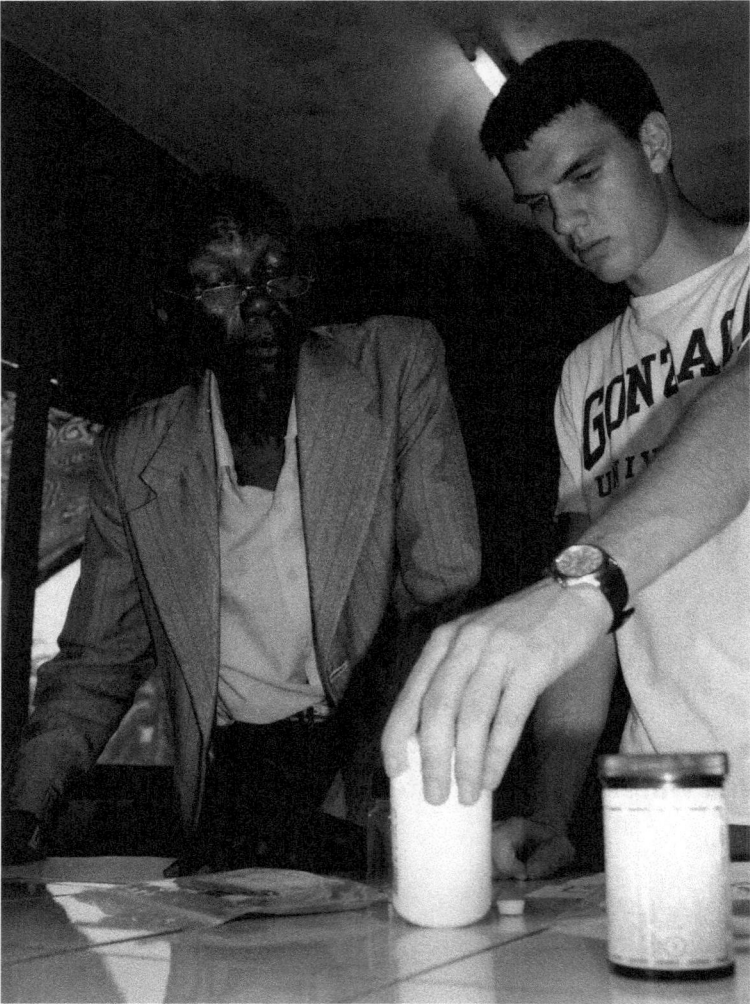

Teaching water analysis methods to Mr. Wilson the microbiologist at the Songhai Center

After testing the water, the next goal was to educate the staff at the Songhai Center about the water quality and our plans to work with them to begin the process of eradicating these preventable diseases. We worked with over 20 people at Songhai, learning how to bridge the language barrier, while teaching English lessons and trying to learn a bit of French and Fonbe, the local tribal language, ourselves.

I was reminded again that you have to use all your communication skills when working with students who may not be fluent in English. We drew, pantomimed, talked around a word, used sounds...all sorts of methods to get our students to understand what we were saying, and for us to understand what they had to say.

The communications portion of the class was thoroughly enjoyable. I'm not going to lie; I was extremely nervous about teaching. I'm an engineer for a reason: I lack people skills. However, with the help of my group, we were able to come up with successful lesson plans.

Teaching for me was the epitome of the sense of exchange and reciprocity we've been talking about. Although I'd like to think that I helped my students build their English vocabulary and confidence, I think that they actually helped me more. Talking to my new friends allowed insights into the Beninese culture I did not get from walking through the neighborhoods of Porto-Novo. I especially loved talking to girls my age who, even though we live far apart and lead completely different lives, I was able to connect with easily.

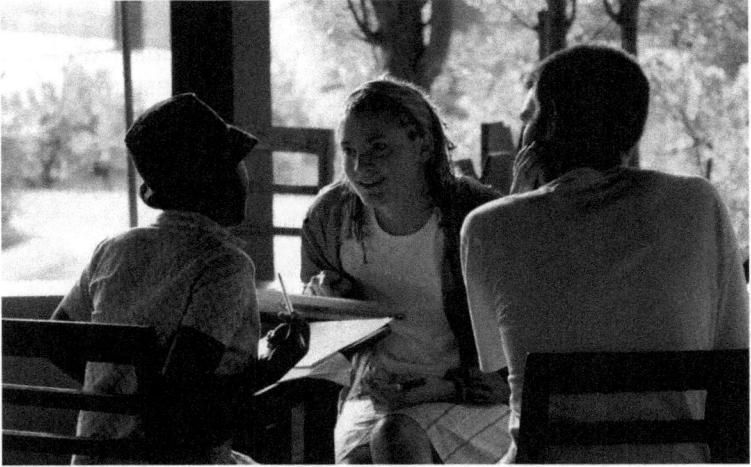

Language lessons with the Songhai staff

In this manner, we were able to start spreading the word, that there is a new game in town and Songhai will have a new tool for development: a simple, cost-effective water filter, which removes disease-causing organisms.

The people of Benin are amazing. I was really worried about the language barrier and having people run out of patience, but that is definitely not the case. The people of Songhai have so much patience and are so open to learning.

Making Water Filters

After explaining to our friends our common goal, to transfer water treatment technology to Songhai, we set about doing just that. Working side by side with our new partners, we got right into the hard work of preparing the clay and sawdust mixture. Both the clay and sawdust were gathered from local sources.

Clay and sawdust are mixed together to form the material that is put into the clay press

The raw materials had to be properly dried and prepared before they could be mixed together with water to form what would eventually become a drinking water filter. We worked with our new friends, Thierry, Michel, Frauque, Raoul and Hacquim to begin forming the drinking water filters.

33

The portable clay press was designed by engineering students at Gonzaga

After the mixture of moist clay and sawdust was prepared, the drinking water filter needed to be formed. A Gonzaga clay press was brought with us on the airplane to teach our partners how to form the filter. Gonzaga engineering students designed this press the previous year.

One thing that really bothered me before coming here was the possibility of the press breaking. I was afraid that it would be extremely difficult to get a new piece that might be needed to repair the press. But this will not be the case; their metal shop is amazing. It's a good thing too because the press was damaged during the flight over.

Beninese ingenuity has surpassed my wildest imagination. We asked for screens to sift sawdust and they put something together in a matter of minutes. We also asked for steel plates for the filters and they cut a bunch of them in just a few hours.

It has also touched me to see how Father Nzamajo's words about Africa sustaining itself have truly come alive. Watching the mechanics work with us on the Filter press has shown me just how true those words are. Everyone up there from Big T to Hakim have impressed me by how fast they learned to use the water filtration technology. More than that, everyone seems to have the Songhai mission deeply rooted in their minds, showing their commitment in everything they do . . .

The Songhai staff with the first ceramic water filters at the Songhai Center

After a long drying period the formed filters were ready to be fired. Of course a kiln had to be constructed for this purpose. Burt Cohen, from Potters Without Borders, was the expert potter who accompanied us to Benin and he constructed the kiln. Burt's presence was made possible by a grant from Rotary International's Spokane club and funds raised by the Gonzaga's Engineers Without Borders club.

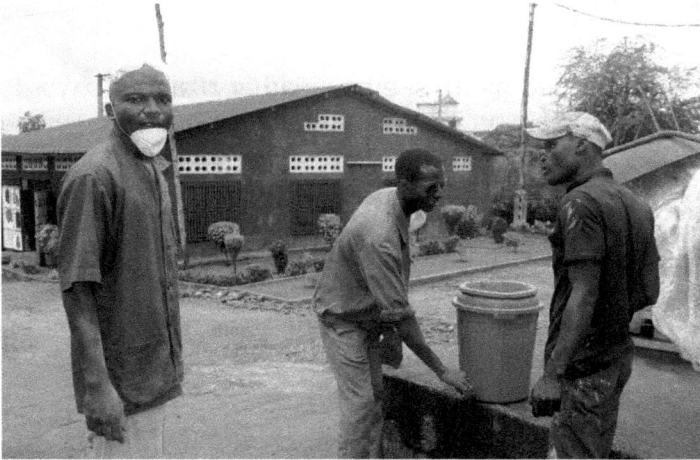

After the filters were fired, Wilson and Thierry tested the filters to ensure that they would indeed remove the pathogens. We were all delighted to hear from Wilson that the filters we made with our partners at Songhai did in fact provide clean water from what had been dirty water, and that the filters are being used at the Songhai Center.

The Songhai Center continues to produce about 25 drinking water filters a day! These filters will be sold to the community at cost, or about 6 US dollars. The filters should provide clean water to the homes and children who use them for at least two years. After two years, the ceramic element may need to be replaced, at a reduced cost of about 3 US dollars.

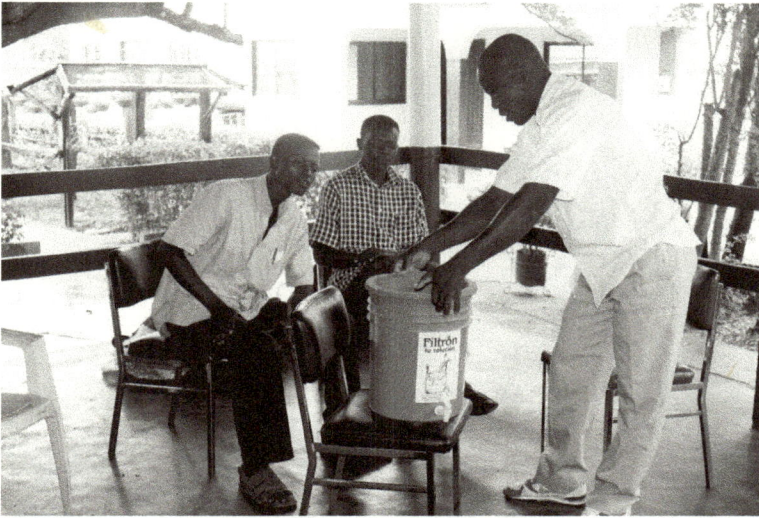

Demonstrating how the filters work

Smiles seem to be everywhere in Benin

The Next Step, into Benin

Even though all the effort on Songhai's part was incredibly nice, there was something unsettling about it as well. Life in Songhai felt separate from what life in Benin as a whole is like. I didn't fly half way around the world to be put up in ritzy touristy accommodations and treated like the wealthy American. I wanted to see and experience real Africa and not just a tourist destination. Well, I got to experience real Africa today.

The world changes once you step beyond the walls of Songhai. You are immediately slapped in the face with REAL Africa. It's enough to make you want to cry and that's not just because the sky is so hazy because of all the pollution. It makes you want to turn right back around and walk back in to the safety and sanctuary that is Songhai.

I want to help them. I want to tell their story to the world.

Benin is like nothing I could ever put into words. This is my first developing world experience; and no matter what you read or see from books or television – you cannot grasp it all until you are finally here.

When we drive around and walk through the neighborhoods, I see many images that break my heart. There is sewage and trash in the streets. People wear filthy, torn clothes. The children's swollen bellies show proof of malnutrition. I know that even if I don't take a single picture, these are the images that will stick with me; these are the things that I will tell people about back at home. However, when I describe these sites, people will inevitably be moved and pity Beninese people. I expect my friends and family to ask something along the lines of, "What should we do to help?" This is where I struggle to find an answer.

I am still trying to process all that I have seen. I knew that there were different lifestyles throughout the world, but I had never seen it first hand. Now that I have, all I can see are the glaring differences between the lifestyles. Perhaps this is only the beginning, to fully understand who I am, I need to get to the point where I see beyond these differences and also see the similarities. This is something that is much, much easier said than accomplished.

Another thing that I have noticed is a feeling of superiority. For the first time I feel treated different because of the color of my skin. I'm not even treated poorly. In fact, those who are black treat me as though I am better than them. They do whatever they can to satisfy us and put us in front of them. I don't like it. The color of your skin shouldn't matter. It should have never mattered.

Our luxury accommodations while staying at the Songhai Center

Following Others' Footsteps

I didn't really know what was going on at first. Brad pulled me out of my group-planning lesson and told me he had somewhere he needed to be at four o'clock. I was supposed to find someone else to drag along who spoke some French or Fon. So, I immediately thought of Alex and together we hurried behind Brad down the street.

We later found out that the previous day he had been walking in the local community and come across a group of small children. The children let him take their picture. Brad had printed out a copy for each child in the picture. So we were off to find the group of kids to deliver the pictures. I thought Brad knew exactly where we were going, but unbeknownst to Alex and I, Mary had led Brad on a rather adventurous journey through the labyrinth of alleys and side streets the day before. So, when it came down to locating the exact spot where the photo was taken, we were pretty much at a loss.

So, we started showing the photo to groups of children along the street hoping that someone would recognize one of the faces in the photo. Sure enough our technique worked. After about the third group of kids, one boy happened to recognize the face of a small boy hidden in the very back of the photograph. He motioned us to follow him since he obviously knew where the little boy lived.

Mind you, this boy that was leading us couldn't be any older than say seven or eight. He led us about ten blocks from where we were to a little alleyway where a group of mothers and their children were sitting on a stoop. We found the little boy in the picture and gave it to him. The smile on his face went from ear to ear. He clutched that photograph in his hand and didn't let it go the whole time we were there. He then ran over to his mother to show her the gift we had just given him. Then the excitement elevated.

Little kids swarmed us. Everyone wanted their photograph taken. It was like at that moment we changed from Yovo with a camera to a friend that cared. We even were invited into their home. I was overwhelmed by their smiles and happiness.

Children in Porto-Novo competing for the best smile

Unfortunately, we had to leave but we promised them that we would return soon with copies of the photos we had taken that day. I thought the previous day was emotional, but today was even better.

We were a little late arriving but as we turned around that corner of that alley, all the kids jumped to their feet from the stoop and came running at us. They had been waiting for us. They knew we were coming and were so excited to see us. It just goes to show the power of a verbal promise. In Benin you are bound by your word. If you say that you will do something or that you will be there you better well do it. I was happy that we could brighten the day for this small family but also a little worried at the same time.

Will the next group of Yovos that come through with cameras return the next day? Or did we just set them up with a false representation that will later cause disappointment?

Statue of King Gliere in Abomey, Benin

Revisiting the Past

Benin itself does not seem as harsh and cold as I first thought it was upon arrival. The people here are much friendlier and accepting of us than I had initially thought. I am still terrified to cross the street though and still feel like 'white people on parade' when we do excursions on foot outside the Center.

We traveled four hours from our home base in Porto-Novo to visit Abomey, the historic capital of the Dahomey Empire. The history of the kings in the museum was very interesting. I did not realize how water-centered Benin's culture was. I found it interesting how in America we offer coffee or tea to the visitors in our homes, while here they offer water first and it is their 'best water' that is offered.

After learning about Benin's history we were able to learn about the culture. I'm beginning to understand why Benin is deeply entrenched in tradition and respect. There is a strong influence from Vodoun (what we call Voodoo) and it was apparent after visiting Ouidah. The whole day brought some interesting experiences really.

We started off with another very long bus drive up to the Songhai Centers in Lakossa. Thank goodness our buses are comfortable! We arrived around lunchtime and ate our sandwiches at the administrative center before driving to see the farm. There we saw crops like rice and corn being grown as well as another large fish hatchery. We met some of the workers and took their picture, leaving a few behind thanks to a Polaroid camera.

Workers shoes in Lakossa, Benin

At the Temple of the Pythons

After visiting Songhai in Lakossa, we drove to Ouidah. Our first stop was at the Temple of the Sacred Pythons, which is located directly across from a Catholic church. We learned a little about the Vodoun religion while we were there. At this temple, snakes were an integral part of the worship. We learned that the temple pythons were let free to go when they needed to feed, and that people from the community brought them back when they were found around the town. Our guide brought a python out that many of us wore around our neck. In the spirit of adventure, I allowed him to put it on me too. It was surprisingly cool to the touch, and a bit heavy, but not too scary. I have lived to tell the tale, although I'm not in too much of a hurry to return any snakes.

Contrary to Hollywood's depiction, Vodoun is practiced for healing many of the diseases caused by poor water. Many people carry the scars under their eyes from traditional healers, who were trying to cure cholera with the Vodoun.

After touring the Temple of the Pythons, we proceeded to the monument of the slave trade. I mostly was overwhelmed by the simplicity of the monument.

It wasn't all that lavish, which ended up speaking volumes. The monument is a carved stone archway that has reliefs depicting the African slaves on either side.

Next to the archway are metal sculptures that are depictions of people in chains. You walk up to the archway on a sandy path, and the arch frames the ocean. It really is quite beautiful.

Memorial marking the point of departure for people sold into slavery from West Africa

The slave memorial was one of the most emotional things I was able to witness during my experience in Africa so far.

To be able to place my feet in the same sand that countless Africans drudged through in shackles filled my soul with sorrow. It was here that they were ripped from their families and homes and marched toward a life of death and pain. These people where no different than me, other than that they were born of a different color. They didn't deserve that fate. As I walked through the arch and watched the carvings of people's backs turn into people's faces, I said a prayer for all those innocent people who were forced aboard a ship at the Point of No Return.

Something else has been bouncing around in my head that Father Joe (Nzamujo) said in Mass yesterday. Summed up, it was about how the slaves were able to continue due to their belief that tomorrow will not be the same. That phrase can mean so many different things to so many different people. It means to me that if today is a good day, enjoy it because it may not last, and if today is a bad day, have hope because tomorrow will be different. I think my understanding and communication of this passage can be so valuable for my future patients that feel fearful, powerless, ill or in pain.

My understanding of this passage can also help me through my rough days, be it personal or professional; hence I am having a bracelet made with the word BELIEVE.

The slave memorial sits on a beautiful site on the shore. I put my toes in the Atlantic with my friend Sarah, but ended up wet to the thigh when I misjudged the wave. After rolling up the pants, I headed down the beach, and joined a Beninian named Coolio dancing to some drummers. I figure when in Africa, you should jump at every opportunity! I had a great time being braver than usual. On the way back to the bus, I stopped at a small craft market to purchase some gifts for friends at home.

One of the things we have noticed as we have ventured into Porto-Novo is that the people we see will not initiate a hello with us Yovos. But when we great them with a bonjour, or a wave, or a smile, the reaction is almost always a huge smile or a greeting back! Though the people we have met in the town can be slow to trust, once you have earned it, the hospitality is amazing.

I understand now why people fall in love with Africa and her people.

Driving to Abomey and Ouidah, you see such beauty and richness here. There may be poverty, and a hard life, but the people are amazing! They live their lives so differently than we do. It is an open living: outdoors, with each other, very little is hidden away. The community supports its members, and works together.

A traditionally carved fishing boat on the beach

The WATER program is about bringing people together to provide a better understanding of the world we all live in

Final Steps in Benin

I'm so glad that I am not leaving Benin with the same impressions I had last week. I was so struck by the sad images I saw – the filthy streets, the malnutritioned children, the houses that were falling apart – that I was focused too much on the differences between Benin and the United States. Now that I've been able to connect with some people, I can see the country for the people. I can think of the health and clean water issues in terms of what individuals would want for themselves and their families.

Being in Africa has already taught me so much about myself. This is the first time I have ever really considered graduate school. This is an amazing story and it needs to be told. If it means going to graduate school to make it happen, it's what I might have to do.

I know that I have learned more from our students than I think we have taught them... The Songhai staff was so observant and intelligent. I know that I will miss them when we return home. I honestly can't believe we are going to be heading back home so soon! I am very excited for our little banquet and to see all of our students together.

The folks of Songhai treated us to a truly amazing goodbye presentation. It was heartwarming to see how truly appreciative our friends were. There was dancing, singing, drumming, and even a feast of traditional African food. It was an amazing thank you and send off on their behalf. It wasn't something I was expecting. What a gift for them to give us.

Working directly with the people of Songhai allowed for some very strong relationships to develop. I really do hate goodbyes, but we can always keep in touch with e-mail, and who knows, perhaps one day I'll be back to visit.

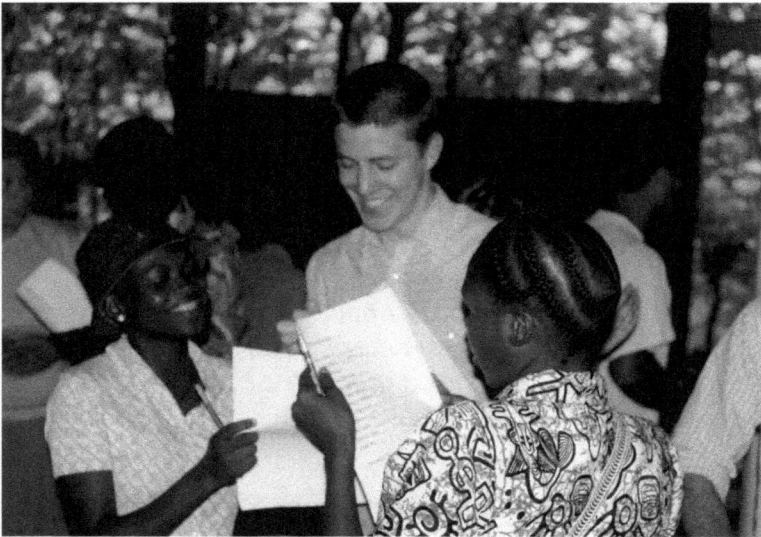

These new relationships will help me stay committed to Benin and help people like Marshal achieve what they want and need. This is all something I never expected to learn from teaching an English class – it is something that will keep me tied to Africa forever.

Reflections

It has finally begun to sink in that only a week ago I was in Benin.

What a trip it was. From the people to the smells to the food, it was all something that can only be understood through experiencing it for myself.

I definitely appreciate everything I have at home so much more, even the little things like drinking out of the faucet. I almost feel guilty being back and knowing that there are families in Africa struggling just to obtain clean water.

I am telling all of my family and friends that the experience was wonderful and sad at the same time.

More than anything else, I've been shocked by how many people seem to brush off the issue of unhealthy water in developing countries. When they hear that we were there to work on a water filter project, they usually ask what I drank while I was there and if I got sick. Then at least half of the people I've talked to mention something about how I would get sick since I'm from the U.S. but that the people living in Benin don't get sick because they've built up enough antibodies. They're surprised every time when I explain that even though they do have some antibodies, they definitely get sick and children live with diarrhea daily.

56

Where do so many Americans get this idea that people in other countries just adapt to living in less sanitary conditions?

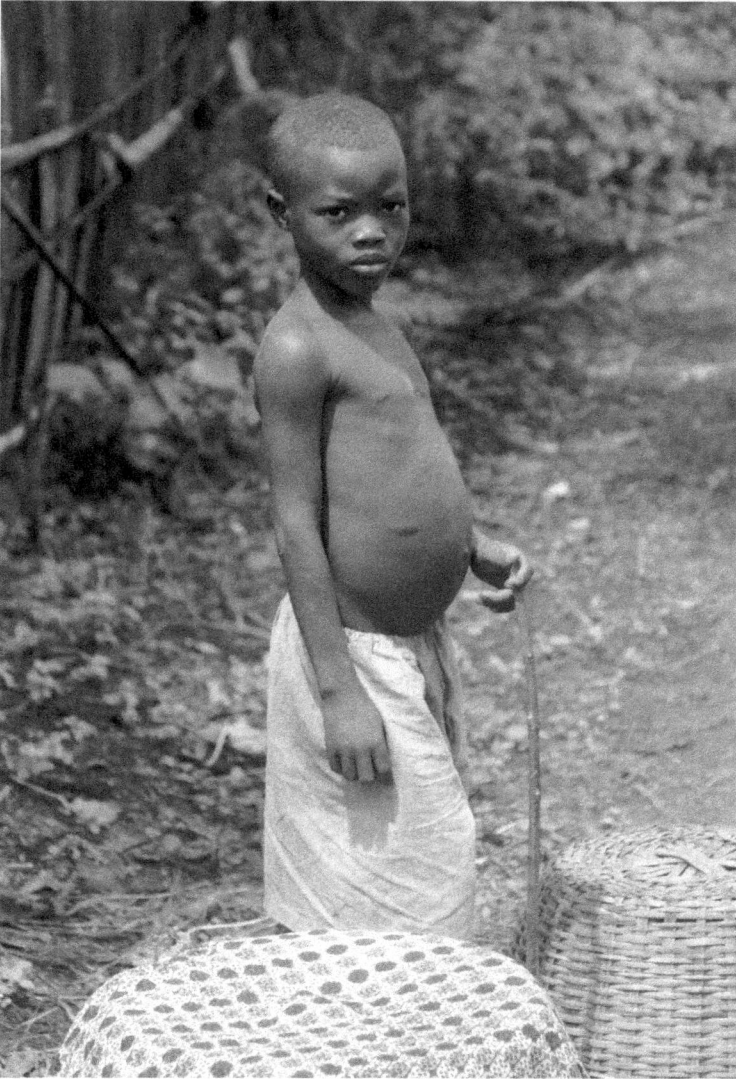

The store at the entrance of the Songhai Center

How am I changing as a result of this experience? I have learned that I can survive 2 weeks worth of cold showers. I can say this is the first time that I have been interested in having as much of a self-sustaining living environment as possible. I am realizing what it is like to truly be a minority. I am realizing what it is like to truly stick out like a sore thumb. I am realizing the value of a double soy latté with toasted marshmallow syrup. I am realizing the value of cheese. I never knew I was such a cheese addict! I am realizing how much Americans rely on modern conveniences; i.e. the store providing food, the faucet providing clean drinking water, the stove providing our means to cook, the microwave providing quick meals, etc. I think we sometimes get so wrapped up in the destination, i.e. quick result, that we do not take the time to enjoy the journey-or even to allow a journey for that matter!

I even wrote about my own petty little issues in my personal journal while we were there. "Here I am crying over spilled milk essentially while I am in a place where people walk everywhere, work for low wages, eat non-nourishing filler foods, and drink dirty water." It has given me a different outlook on my own priorities.

My pre-trip metaphor held up: I do feel like I was the gullible and naïve child entering school for the first time. I do feel as though the 'picture books' did not do our experience justice. How could they? In all actuality, I may feel more like the naïve child now than I did prior to leaving. I heard one of the students say, "You don't know what you don't know." I am now more aware of what I do not know. Now I have the choice on how I want to deal with that knowledge of ignorance. Awareness is half the battle though and the lessons I have learned from this trip are still very fresh.

Though I am so far away, I have had an experience that has taught me things and shown me things I will never forget. And now, I have a choice – what to do with that experience. I am not sure what direction it will take me, but one thing is for sure, I will not let my lessons learned there slip away from me.

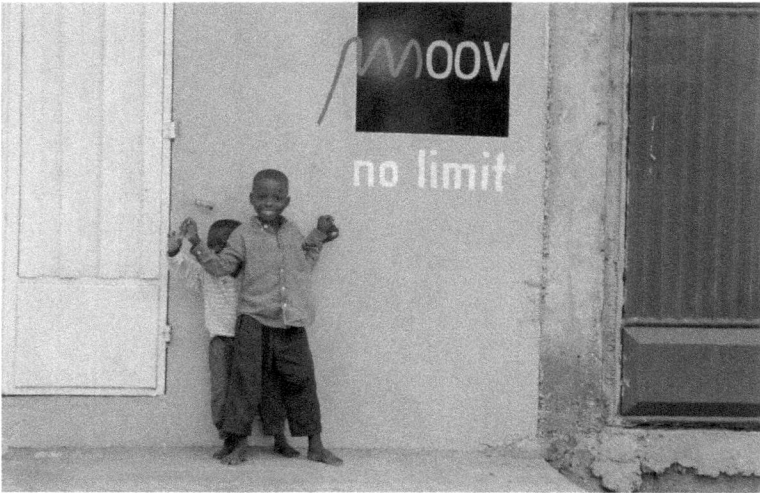

New Steps and New Directions

I do know that I have found my calling. It is such a great feeling, once you figure out what your passion is and just knowing that you are at the culmination of all your education and all your growing up. It is an intense feeling to wake up in the morning and knowing in your heart that you are exactly where you are supposed to be, geographically, emotionally, spiritually, and in life in general. I know that this is the field in which my skills and all the gifts that God has given can be used to their fullest and I can be all that I was meant to be. Sorry that's only a little cliché to invoke the old army motto but it feels strangely appropriate. This is all very exciting and scary, but I just know that what's right is right and that is what Africa has given me. This trip has given me the motivation, courage, and knowledge that I will put to good use for many years to come.

I want to also thank all of you for joining me on this journey and for all the support you have given me through it all. It really has been a life changing experience, and I owe a huge debt to all of my family and friends who have supported me financially, spiritually and emotionally as I've gone through the process. You can't know how much I appreciate everything!

Organizers of the WATER program from the Songhai Center and Gonzaga University

I wish I could tell you that our work was completed, but Father Nzamujo has other plans for us! If possible he would like Gonzaga's WATER team to return next year to set up another drinking water filter production center in Savalou, a more remote town in Benin. I was also fortunate enough while I was there to meet the owner of Ecobank, the largest bank in Benin while we were at the Songhai Center. He was visiting to determine the site for a new health clinic at Songhai! By next year, they will have a health clinic building, but no trained health care providers, unless we are willing to return to provide basic first aid training. In addition, Gonzaga business students are developing a marketing plan to ensure the new filters find their way into the homes of the people in Benin. Gonzaga engineers are working to make the filters work even better and cost even less.

In upcoming years, health assessments, educational materials, and technical data will be used as metrics to measure the success of the project. As part of the on-going collaboration between the Songhai Center and Gonzaga University, an epidemiological study will assess the implementation process. This study is anticipated to last four or more years in Benin. Other potential WATER sites are being investigated in Kenya, Tanzania, and Zambia.

However, the program is difficult to sustain without outside support. To date the program has been entirely self-supported, funded by grants from the US EPA and small private individual donations. The faculty donated significant amounts of their time in order for the program to exist. We hope to be able to continue the WATER program at some point in the future, pending the outcome of future grants or alternative support.

I would request that we all thoughtfully consider the potential this and similar experiences have on the students and people we serve. Perhaps we might consider what one student shared in their journal:

"I think that in the future I will look back at this WATER class as one of the most significant choices I have made in my life, and I know that in some way just the act of participating will have some effect on all the things I do in life from here on."

The WATER teams from Gonzaga traveled halfway around the world to transfer technology, provide education, and receive from our new friends a new way to look at our world and theirs.

We All Have Gifts

By Nadia L. Warren, MA-TESL

Although the world appears to be smaller due to technology, in many ways, we are very isolated. The key to breaking through the isolation is communication and education. Interdisciplinary projects such as the WATER Project in Benin, West Africa in August 2007, remind us that we are not islands, separated from each other by our respective "specialties", but are instead, parts of a whole, each with a valuable contribution to make, each with something to learn and something to give.

Exploring knowledge outside the boundaries of our own disciplines increases our usefulness. With the walls down, we are also more inclined to establish relationships with people, which, in turn informs us of their needs, which then makes us more effective. The "stretching" process keeps us humble and teachable. I believe it is safe to say that all of us who participated in the WATER Project were very cognizant of the fact that we were both teachers and students ourselves and sometimes the line between the two roles was very blurred. We went to Benin to help equip people with a vital tool in reclaiming their health and futures, but I think it safe to say that we spent the better part of our time there very much aware that we were learning at least as much as we were teaching! The process began with the two and a half months of intensive study of epidemiology, engineering and teaching that preceded our departure, and continued as we had to put feet to the academics in practical application with very real people and very real situations. We were challenged with finding culturally relevant means to use existing knowledge and understanding on which to build new knowledge.

During our two weeks at the Songhai Center, we made Filtrons, assessed sanitation and hygiene needs, and taught English as a Foreign Language to Songhai staff. Some worked in the Songhai Center Lab, teaching how to test for contaminants in water while others went out into the community and asked permission to draw water from various wells so it could be tested. We talked to people who depend every day on these wells for water to drink, bathe in, and cook. It is all they have and although most know that the water is full of deadly organisms, what other choices are there? Bottled water is available but for many, this is not an affordable expense.

The faces of the people reflect pride, resignation, stoicism, hope, frustration and more. They know, to varying degrees of understanding, what their circumstances are, but the sanitation, hygiene and water problems are so all encompassing, many simply resign themselves to the situation, and by doing so, perpetuate it.

As the Songhai Center has gained greater recognition both nationally and internationally, the number of visitors has increased tremendously and the need to communicate with them has become essential to the people who work there. The request was for English lessons tailored to specific areas within the Songhai Center. The restaurant staff, first aid workers, shop mechanics and engineers, and administrative personnel all have opportunities to interact with multinational visitors. The challenge for us was to tailor the English lessons to meet each of these groups needs, but since the classes were grouped by skill level, not job description, our interdisciplinary group of teachers taught interdisciplinary classes! It worked well as the students were very teachable and eager to learn, exhibiting a very healthy competitive streak! They all learned vocabulary specific to their own and others' areas of expertise and were in a sense, cross-trained.

The lessons continued outside of the classroom as we interacted and began building relationships with people inside and outside of the Songhai Center. What had been learned in class was practiced and expanded upon with conversations outside the classroom that served to teach (on both sides) and build ties. These ties were on a personal level and also continued building on the foundation of the relationship between Gonzaga University and the Songhai Center that was begun four years ago.

Our group was very diverse, consisting of engineers, nurses and teachers. For some, the trip to Benin was the first time outside of the U.S. We all gained a great deal of new knowledge academically and personally. We learned new ways to teach English by making it relevant to the needs of the students and had a refresher course in appreciating the difficulties of learning another language. It is easy to see how communication, education and relationship go very far in transforming lives and attitudes. Talk to any of us who went to Benin and you will understand that lives were changed on both sides of the Atlantic.

An African Perspective

By Gilbert Nalelia

Water is life. The water on our planet earth covers about 70%. About 2% of the water is fresh water and the remaining is salty water that is useless for human consumption. Only half of the fresh water is available for use by the ever-increasing population of the world, which is approaching 6.7 billion. A third of the nations on our shrinking planet, mostly in the developing world, suffer from water scarcity and the associated stress this scarcity places on societies. My home area of Kitale, Kenya shares in this suffering.

Water is critical for human survival and health, and to all aspects of a society's social and economic fabric. Its availability in both quantity and quality has significant bearing on commercial productivity and economic growth. As the population of the world grows with its associated increased levels of economic development, human needs for water march in parallel as increased pressures are placed on the resource hungry sectors of agriculture, industry, energy and environmental management. As a result of these increased uses of water, a completely new approach to water management (termed integrated water resource management) has evolved. Integrated water resource management embodies a concept which takes into account social, economic, environmental and technical dimensions of water usage and provides tools to regulate the many demands placed on this shrinking resource. One of these tools prescribes that decisions should be based on comprehensive evaluation of weighted scores assigned to each "competitor" in the resource queue.

Multiple reuses of water in different sectors and at varying stages has taken center stage as compared to the more simplistic past where there was a more defined separation of water source/use for agriculture, industry and municipal needs. This concept is illustrated below and clearly differentiates the Sectoral divisions.

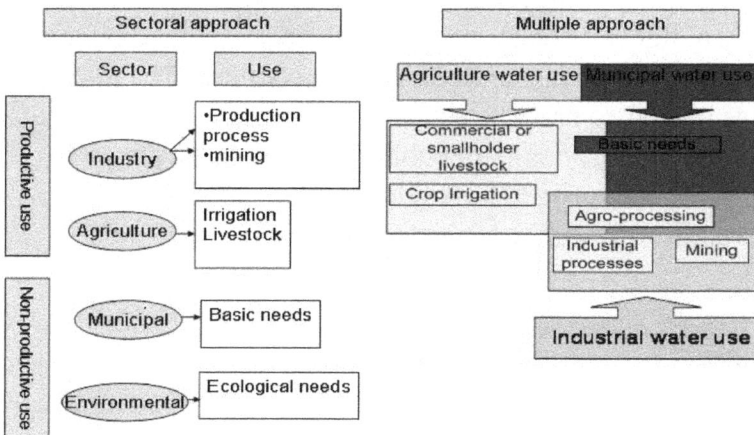

Sectored and multiple use of water

Over the last 50 years, the population of the world has doubled and its water use has tripled. In many developing countries, water availability (both quantitative and qualitative) is emerging as a major developmental challenge. It is estimated that of the 800 million people living in Africa, more than 300 million live in a water-scarce environment. Kenya, being a member of the family of developing nations, is faced with this developmental challenge both in the major cities and the rural areas. Water quality and availability is an accelerating concern due to the pressures of expanding urban settlement, increased industrialization and agricultural activity. Unfortunately, my country's infrastructure is not adequately equipped to meet this challenge.

The semi-arid and arid areas of Kenya are a perfect case study of the typical use of water in most of sub-Saharan Africa. Kenyan communities are highly influenced by the availability and use of limited water sources. The impact this has on communities is far greater than the traditional "public health" numbers published by the Kenyan government. The following categories of water use by rural communities can be identified as:

1. Water for domestic needs: These uses are focused on simple survival, providing water for cooking, drinking, sanitation and hygiene which have significant impact on health and well being.

Compounding the challenges for domestic needs is the statistic published by the World Health Organization (WHO) which documents that 3.4 million people die annually of water-borne related diseases throughout the world. The majority of these victims are children under five years of age who perish from dysentery or diarrhea.

2. Water for productive activities: These uses impact farming and other income generating activities that use water in their processes. Examples include subsistence farming and commercial farming. Other activities might include car washing, hair salons, food preparation, etc.

3. Water for other uses: These uses include cultural activities and religious ceremonies.

I was born in the small community of Arpollo nestled against the eastern base of the Cherengani range in the Rift Valley. In Arpollo one experiences the rawness of nature's basic call for water and the stark fact that the water must be carried to the community from a river five miles to the east out on the Rift Valley floor. This Rift Valley river water is subject to all of the uncertainties found in any Kenya river which hosts the entire range of contaminant sources afflicting human health.

Due to rural-urban migration my parents moved and settled in Kapenguria, situated 20 miles north east of Kitale where they have lived for the past 22 years. Kapenguria has a small municipal water system that is over 20 years old. The district department of The Ministry of Water and Irrigation is in charge of operation and maintenance of the water system. There are two tanks in the system. The primary tank is located 15 yards from the river, the source of the water. Water from the river is lifted into the first tank by an electric pump. Basic sediment removal and screening is performed before it is pumped up to the secondary tank. The secondary tank is situated on the ground about 1000 yards from the primary tank on a slope approximately 30 meters higher than the primary tank. In the secondary tank, the water is chlorinated before being distributed to homes, restaurants, clinics and the hospital.

Since commissioning the water system, little effort has been made to maintain the system. Over time water access has diminished due to customer billing, illegal connection to the distribution lines and general political/departmental corruption. Resultantly, there are insufficient funds to repair primary infrastructure problems such as broken distribution lines and broken electric pumps. Currently the system delivers water to only about 20% of its customers. Electricity is erratic causing the pumps to fail and limiting water access. A simple water tower that uses gravity would deliver water at a constant pressure into the network. Unfortunately, the system was flawed from its inception by the absence of a such a tower.

Over the last couple of years, my family has resorted to fetching water from the local river after the pipes connecting our home to the distribution system rusted because of age. Complaints to the relevant authority have fallen on deaf ears or with empty promises of repair "coming soon." This has prompted my family to look for other sources of clean water. Fetching water from the river is unsafe, time consuming and unreliable. A well with a depth of a least 50 feet (bore hole in Kenya) is the only viable option for securing safe water.

Similar water shortages exist in Kitale, an agricultural town in western Kenya situated between Mount Elgon and the Cherengani Hills at an elevation of approximately 6200 feet. It is known as the "Bread Basket of Kenya." The population of Kitale and its surrounding communities is approximately 150,000 people. As is the case for the majority of Kenya, the Kitale area regularly experiences water scarcity. The Kitale Municipal Council is mandated to manage the distribution of water within its 3 kilometer radius of operation. The Council's department of water and sanitation is in charge of operations, maintenance and improvement of the public water supply system.

The sources of raw water are the Koitobos river and Nzoia river, each of which have separate treatment works. The Koitobos facility was initially designed with a capacity of supply 2,600m³ per day. The facility supplied by the Nzoia river has a capacity of 10,000m³ per day. The two sources provide a combined capacity of 12,600m³ per day. The daily commercial use of water within Kitale municipality is about 5,030m³, while the daily domestic and livestock consumption is about 12,150m³ and 200m³ respectively.

This leaves a shortfall demand for water of 4,780m³ per day causing water rationing. The demand for water of 17,380m³ per day can only be achieved through rehabilitation, improvement and expansion of the existing schemes.

Water remains a challenge throughout sub-Saharan Africa. Water access must be made a priority in Benin, Kenya and other developing areas of the World.

Reference:

1. Economic Commission of Africa, WATER IN AFRICA Management Options to Enhance Survival and Growth. September, 2006.

2. http://www.who.int/en/

3. The Ministry of Water and Irrigation, THE NATIONAL WATER SERVICES STRATEGY (NWSS), Period 2007-1015. June, 2007.

WATER Impacts

By Sushil Shenoy

The West African Technology, Education, and Reciprocity (WATER) program that I was a part of at Gonzaga University turned out to be a vital experience for me. It impacted my life in ways that I had never thought possible. I was fortunate to have already traveled the world rather extensively before I entered college. By the age of eighteen, I had lived in Europe for two years and spent over a month in Asia. As a result, I had a fairly good idea of how the rest of the world operated.

The key thing that the WATER program provided for me that I lacked was a trained mind. I had several experiences while visiting or living in other countries; but I had no idea of how to solve what is considered a basic problem here in the US, access to clean drinking water. How can one get drinking water in sufficient quantities and at a sufficient quality to improve people's quality of life? Normally in the US, it is simply matter of drilling a well or building a water treatment facility. The vast majority of the world either does not have a well-drilling rig or the money to build and operate a water treatment facility. By providing me with the education and practical experience, the WATER program has trained my mind to approach these types of problems.

Since most of the world cannot afford drilling equipment or treatment plants, one has to look at other ways of providing fresh water in a way that is appropriate for the location under consideration. It is important to note that it can often be a benefit for other countries to be unable to afford the solutions we employ in the US. In fact, the solution in that other part of the world will almost always be able to provide the same quality of water in a much more sustainable fashion. The key is using the appropriate solution for the given location. For example, the solution for one part of Africa is not the solution for another part of Africa. The WATER program taught me to look at the conditions at any given location in the world before deciding how to provide clean water.

When I was in the WATER program, we looked at the best way to provide water for people in Benin, West Africa. We began by gathering information on the different soil types, the rainfall patterns, and as much cultural information that we could obtain. Then we looked at all the different methods for cleaning water. Some of the methods we explored are solar disinfection, bio-sand filters, clay filters, and other methods like Lifestraw. Additionally, we made a matrix with each method on one side and the local conditions on the other. An example of a local condition is the availability of clay or how far someone would have to walk to get to the water source. We determined the best solution for Benin by using this matrix and advice from people who had either been to Benin or had been somewhere close to Benin.

The solution we decided to use is called a Filtron, and it is simply a clay filter that uses the microscopic holes of the clay to only allow water threw it. The practical experience of taking our solution to Benin and implementing it instilled in me the importance of the process we had gone through to create the final solution. If we arrived in Benin with a solution requiring pumps, the people there would not have had the parts to maintain the pumps. Once the pump broke, the solution would have been useless. Thinking about considerations such as these is the crucial training the WATER program provided me with. For those who have never been to another country, the benefit of this program is even greater because it will expose you to the rest of the world. It will broaden your horizons and help you think outside the box. There are parts of the world that do things very differently than we do in the US, but their methods are just as effective, if not more effective at providing for a person's needs.

I am currently a graduate student at Virginia Tech. Our School of Construction is trying to start a program for sending one or two graduate students from the School to different parts of the world. The School will provide teams from different universities a project manager type service. For example, a construction graduate student would join a team of political science students from Notre Dame (or anywhere) and go do a service project like the one I did in Benin in Vietnam or some other place. The graduate student from Tech would need experience in project planning and execution in different parts of the world. My education at Gonzaga and the training the WATER program gave me, has uniquely qualified me to help with starting this program. The program is in its infancy, and I have been asked to go ahead of a team of students from Virginia Tech to prepare a project we are planning to do in Belize later this month. I feel my leadership position in this project is primarily due to my experience with WATER. A similar opportunity is available for people interested in medicine, education, or other disciplines. For this reason, I recommend the WATER program since it has been a vital experience in my life.

I cannot say for certain yet, but I believe my involvement in this program has changed the course of my life. I would not be at Virginia Tech had it not been for WATER. My goals for graduate school changed for me after my experiences at Gonzaga. I now looked for a school that was focused on sustainability, and that wanted to help other people in the world. For anyone in an undergraduate program at Gonzaga, I urge them to consider going on to do graduate work at another institution because this is where you can begin to tackle real world problems. For those interested in helping others like I am, I would also strongly recommend the WATER program. You never know where it may lead......

The Water Crisis and Implementation Strategies

By Jessica J. T. Oddo, RN, BSN

A silent humanitarian crisis kills 3900 children everyday (Bartram, 2005). The root of this unrelenting catastrophe lies in these plain, grim facts: four of every ten people in the world do not have access to even a simple pit latrine; and nearly two in ten have no source of safe drinking water (Bartram, 2005). Far more people endure the largely preventable effects of poor sanitation and water supply than are affected by war, terrorism, and weapons of mass destruction combined. Yet those other issues capture the public and political imagination, and resources, in a way that water and sanitation issues do not. Most people find it hard to imagine defecating daily in plastic bags, buckets, open pits and public areas because there is no alternative; or cannot relate to the everyday life of the 1.1 billion people without access to a protected well or spring within reasonable walking distance of their homes (Bartram, 2005).

Water is fundamental for life and health. The human right to water is a pre-requisite to the realization of all other human rights. The poverty of a large amount of the world's population is both a symptom and a cause of the water crisis affecting the poorest populations in sickness, lost educational and employment opportunities, and for a staggeringly large number of people, early death (Gleick, 2002). Water-related diseases are among the most common causes of illness and death in poor developing countries (United Nation, 2003).   The sad fact is that this disease burden is preventable. In developing countries with poor water and sanitation systems, life expectancy is far lower than in industrialized countries. The causes of deaths are also quite different; infectious diseases account for more than 40 percent of deaths in developing countries, whereas industrialized nations deaths are related to chronic disease and cancer (Kitawaki, 2002).

Recently, water has become an increasingly important issue in Benin. Benin is located in West Africa and is one of the poorest countries in the world. Although it profits from oil revenue, this wealth is concentrated in the hands of a few. Most people rely on subsistence farming for their livelihood. The health situation of the country is greatly jeopardized by access to water. The lack of access to clean water forces many to drink unsafe water which aggravates existing health problems by causing and spreading disease. The following are results of this problem: life expectancy in Benin is 53 years, 63% of the population has access to safe water and only 23% to adequate sanitation facilities. Under 5 years of age the mortality rate is 154/1,000 live births and net primary school enrollment/attendance is 54% (World Bank 2004).

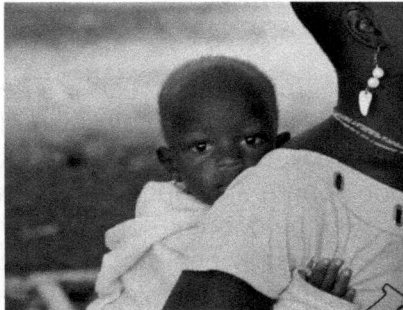

In recent decades, a consensus developed that the key factors for the prevention of diarrhea are sanitation, personal hygiene, availability of water, and good quality drinking water (Jensen et al., 2004). Water sources are often located some distance from the home requiring collection and transport from the source and subsequent storage of water within the household. If the water from the collection site is clean, it can easily become contaminated in transport (Wright, Gundry, Conroy, 2004). This is an aspect that makes the Ceramic Water Purifier (CWP) important as a point-of use source. Studies have shown that improving the microbiological quality of household water by on-site or point-of-use treatment and safe storage in improved vessels reduces diarrheal and other waterborne diseases in communities and households of developing countries (Wright, Gundry, Conroy, 2004).

In Benin, people derive their water from a variety of sources: direct withdrawal from small ponds or streams, traditional shallow wells constructed locally (water in these wells is often not safe to drink), and modern wells that are filled with concrete in order to prevent outside contamination (Heidecke, 2006). Many water sources in Benin are threatened by contamination related to high levels of groundwater extraction from city wells near the coast increasing the threat of saltwater intrusion, high amounts of fertilizers and pesticides used in agricultural production contaminate resources, as well as, the lack of sanitation systems (Heidecke, 2006). Human wastes are managed primarily through pit latrines and open fields; there are no community waste areas. Garbage is disposed of outside the walls of family compounds and in lakes or swamps; often the same places are used for water sources. Some, but not all, domestic animals roam freely about town. Water samples tested by Gonzaga students showed a high level of bacterial contamination. The presence of fecal coliforms indicated that the water sources were contaminated with sewage (Striebig, 2005).

The following studies are evaluation or survey designed research. Evaluation research is an applied form of research that involves finding out how well a specific program, practice, procedure, or policy is working. Various schools of thought have developed concerning the conduct of evaluation research. The traditional strategy for evaluation consists of four broad phases: determining program objectives, developing a means to measure the attainment of those objectives, collecting the data, and interpreting the data in terms of objectives. An alternative evaluation model and one that is most useful to this synthesis is the goal-free approach (Polit, & Beck, 2004). Proponents of this model argue that programs may have a number of consequences besides accomplishing their official objectives, and the classic model is handicapped by its inability to investigate these other effects. Its goal is to describe the repercussions of a program in the absence of a practice on various components of the overall system. Most of the studies reviewed have completed surveys. There is not an established format for reporting a survey's results. A formal report usually contains contextual information, a literature review, a presentation of the research question under investigation, information on survey participants, a section explaining how the survey was conducted, the survey instrument itself, a presentation of the quantified results, and a discussion of the results (Polit, & Beck, 2004).

Lantagne (2002) investigated the Potters for Peace colloidal silver impregnated ceramic filter for intrinsic effectiveness and field performance in rural Nicaragua over a period of three weeks. This study was funded by USAID. Surveys were administered during unannounced visits to thirty-three families using the filter in seven geographically diverse communities of Nicaragua and water quality analyses of pre- and post-filtration water was completed at these sites. The goals of the survey were to determine factors that correlated with correct filter use and to gain understanding of the water situation in each home. Laboratory results agreed with historical data and showed the CWP is capable of removing 100 percent of bacterial indictors of disease-causing organisms. However, research in homes indicates that an educational component including safe storage, cleaning procedures, and follow-up visits is necessary to ensure the intrinsic effectiveness is matched in the field. 73 percent of homes were using it at the time of the unannounced visit and usage varied by community from 30-100 percent. Higher usage rates were found to be directly related to follow-up visits by a staff member or community leader associated with the sponsoring NGO. Usage rates were significantly lower where no follow-up occurred.

Nicaragua is similar to Benin in that only 59% of the rural population has access to safe water. In this study women and children were the primary collectors of water. Average family size was 5.7. All families cleaned their filter at a minimum of once monthly but cleaning technique varied largely, and no family scrubbed the filter hard enough to remove build-up of suspended solids which greatly affected filtration rate.

The apparently contradictory results between field and laboratory testing were found and indicate incorrect usage of the filters. Follow-up education appeared to be critical to the correct and continued use of this system. The most common problems observed were the breakage of the ceramic filter when trying to clean and insufficient flow rate. This shows that proper education on cleaning of the CWP is crucial to continued effectiveness. This study recommends follow-up, development of an educational brochure, and a cleaning kit to accompany the filters. The weakness of this study is that the size of the sample is not sufficient to infer conclusively the cause of the disparity in results.

Eun Young Hwang (2003) monitored the point-of-use ceramic water filter for six months in the field in San Francisco Libre, Nicaragua through Massachusetts Institute of Technology. Eun Young Hwang used Lantagne's 2002 study as a starting point to design the study. This study surveyed 100 households monthly for six months. The 100 households were randomly selected out of 2000 families who had received a CWP. These households received the filter 2 months prior to the start of the study. The objectives of this study were to assess performance of the filters by monitoring flow rate, microbiological contamination before and after filtration, and to identify the variables of contamination, also considering cultural hygiene habits that Lantagne (2002) thought affected field performance of the filters. Both analytical and surveys were used to study flow-rate, microbiological removal and user acceptance of the CWP. Results for the six month monitoring showed that an average of 80.4% of the sample population had safely filtered water and filtration rate averaged 1.7L/hr. Recontamination of the filtered water due to contaminated receptacles was in 33% of the 20% of cases that did not pass as safely filtered water. Users also complained of the small capacity of the filter. Filter breakage was appreciable, since 15% of the filters broke by the end of the study.

This sample population again was similar to the situation in Benin. The average monthly salary in the sample was US$100 monthly, females were mainly in charge of supplying water, the average household size was 4.8, and the literacy rate was 88%. Most houses had dirt floors and firewood kitchens, children went barefoot; they shared living space with domestic animals (i.e. chicken, pigs, and goats).

The study cites its biggest weakness as communication with the field technicians that were hired to help carry out the study. It felt that questions may need to be raised to the validity of some results related to the technicians' lack of laboratory knowledge, previous health or science background. The study proved a need for continuous follow-up after the introduction of the CWP, and its biggest recommendation was careful selection of the follow-up staff. Ideally it is recommended that they have basic science, engineering, or public health background, and have local experience. They should also be involved in all stages of the CWP process to gain a strong understanding of it. The study also found that more education was needed for cleaning the filter and ways water becomes contaminated after filtration. It found that the follow-up and continued education should be aimed at women and that live demonstrations were most affective due to the variable literacy rate. If labels and brochures are designed, they should be attractive and easy to read for children because they are often the ones able to read; then the information is dispersed to their mothers.

Roberts (2003) conducted CWP field tests in Cambodia sponsored by the Health and Nutrition Initiatives Fund supported by the Canadian International Development Agency and the Ministry of Health of the Kingdom of Cambodia. 1000 CWPs were distributed to twelve rural villages to test their performance under household conditions. All recipients were interviewed prior to receiving their CWP and three months after CWP delivery to assess water-related expenses, adequacy of water volume, compliance with recommended hygiene practices, and user satisfaction. There was also a subset of recipients (n=101) that were interviewed to determine the impact on CWP-users after one-year of use to a control group (n=100) that did not have CWPs. This survey measured incidence of diarrhea, time and expense savings, and compliance with recommended hygiene practices.

This study found that the type of benefits experienced by CWP users depended in large part on their water treatment practices prior to receiving the CWP. Households that boiled their water saved time and expense related to water boiling. 69% almost always boiled their water prior to the CWP. Almost all stopped this practice after using the CWP. 89% of the 69% collected their fire wood themselves and saved 22 hours per month in time spent gathering wood and boiling water. The other 11% purchased the wood and saved on average of about $1.40 per month in wood expenses and about 16 hours per month boiling. The household surveys were not able to determine if using the CWP resulted in health improvements because they were already boiling their drinking water. The households that did not boil their water prior to CWP use did not save on money or time but did show significant health improvements. When compared to non-boiling, non-CWP households, CWP users had 17% more households reporting no diarrhea, which meant a savings on diarrhea treatment and fewer work/school days missed. It was concluded that CWPs would pay for themselves in about six months. Most of these benefits affect women and improve their situation because they are again the ones responsible for water collection and caring for the sick. In Cambodia, women are usually the managers of household expenses so they benefit directly from money saved on purchases of water, firewood, and medications.

After one year, about 20% of recipients had stopped using the CWP regularly. Half of those were due to a broken filter. Because of this, the cleaning recommendation was revised from twice weekly to once a month or when the filtration rate slows. Breakage was thought most often to occur when the filter was being handled for cleaning. This was a well written and designed study. It was thought that the motivation of this population was high because they already took step to decontaminate their drinking water and understood what the CWP meant to savings of time and money.

As the Cambodia study showed, the CWPs have a higher success rate when people are already changing hygiene behaviors in relation to water supply and conciseness of ways water can become contaminated. Other researchers examined factors that influenced behavior change in developing countries. The common variables are: poverty limiting access, cultures and customs limiting acceptance, poor infrastructure limiting access and communication, and poor education again being a barrier to access and communication (Thesenvitz, 2000). Certain trigger events also influence behavior change like the onset of the rainy season (which increases the perceived risk of diarrhea), the introduction of a newborn child into the household (which induces protective instincts of mothers), or the need to care for a sick person (Thesenvitz, 2000).

Quick (2003) studied behaviors change techniques in three different African countries, Zambia, Madagascar, and Kenya, that have been used to motivate the adoption of a safe water intervention at the household level to prevent diarrheal diseases. Methods of behavioral change used were social marketing, motivational interviewing and community mobilization. Social marketing is the use of marketing to promote socially useful products in order to change behavior through generation of demands for products. There are four 'p's in social marketing: 'product' high quality and attractive, 'price' affordable and permits partial cost recovery, 'promotion' through education, information, and communication to generate demand, and 'placement' which is wide distribution for easy access. Motivational interviewing is theory-based and incorporates decision theory, motivational psychology, and the stages of change theory. It involves the use of simple counseling techniques, including listening, reflecting back certain themes, and eliciting from the client their own arguments for change. This is done so the client realizes the need for change. Community mobilization involves training community members in the technology and reasons for use. There is also active community participation in research, planning, implementation, and monitoring so that the community develops a commitment to the project and a sense of ownership.

In Zambia, social marketing was used to implement a safe water system in 100 randomly selected households and another 100 households also received motivational interviewing. In Madagascar, the same safe water system was implemented in 100 randomly selected households using social marketing and another 100 households also received community mobilization. In rural Western Kenya, social marketing and community mobilization were used to implement a safe water system in 100 randomly selected households. Results were gathered via surveys and water quality testing. In Zambia, three months after the project lunch only 14% of the social marketing only group had adopted the safe water system compared to the 78% in the social marketing and motivational interviewing group. In Madagascar, three months after launch only 11% of the social marketing only group had adopted the safe water system compared to 20% in the social marketing and community mobilization. These results were felt to be flawed because the launch date was moved up 4 months from the planned date due to a cholera outbreak so careful preparation was not followed as much as in Zambia. Madagascar was then also hit by three cyclones. In Western Kenya, 37% of the households adopted the safe water system. The conclusions was social marketing is an effective tool for disseminating product awareness and motivating those individuals who are already hygiene conscious. Motivational interviewing and community mobilization is needed to prod the skeptics or cynics to consider product adoption and thereby enhance the effect of social marketing.

Thevos (2000) also conducted three field studies in Zambia comparing educational and motivational approaches to behavior change concerning safe water. The effectiveness of motivational interviewing was compared with the standard practice of health education alone in initiating and sustaining safe water treatment and storage behavior among community residents in field trials one and two. In field trial three, motivational interviewing was compared with social marketing. Each field trial had a comparison group. Data was again collected through community surveys prior to local health promoter training and a follow-up. The sample size was 796 from low socioeconomic status peri-urban communities. Local volunteer health promoters from communities were trained in adaptation of motivational interviewing for safe water treatment and storage. All health promoters received instruction in the causes and prevention of diarrhea. Health promoters in the experimental groups received motivational interviewing training.

In field trial 1, there was no statistically significant data to show a difference between the comparison and experimental groups in the eight week trial period. In field trial 2, a significantly greater difference in safe water usage was observed in the motivational interviewing group, 71% more than the education alone group. In field trial 3, the incidence of safe water use increased from 1% at base line to 65% four months later. Similarly, significant increases in knowledge about diarrhea were also found. In the comparison areas, there were no significant differences between the baseline and follow-up data on the measures.

This is a strong, well executed study. Much attention was paid to design, and procedures were followed systematically. There was also much attention paid to communication which had been lacking in the other studies. This study appears to demonstrate the strong potential of motivational interviewing to enhance public health initiatives in developing countries. This could be in part related to the importance of interpersonal communication, as opposed to the exclusive use of mass media. It is a method of respectful interpersonal communication about behavior.

This study can relate to the promotion of the CWP in Benin by learning from one of the first social marketing programs completed in sub-Saharan Africa which is where Benin is located. The lessons taken from this study could be: attention to use of multi-media, clarity of message being sent as not to confuse the issue, and increase the number of languages being used to promote the CWP. In Benin, French is the official language, but there are five other local languages; in the rural areas they often can only speak one of the local languages. Also, like Benin the literacy rate is low at 65%. This must be kept in mind when designing promotions, as well as, the target audience for the CWP would be women so one must find the type of media most engaging to them.

Sumaya (2006) researched health promotion in Texas Colonias using community health workers. Following the 1993 passage of the North American Free Trade Agreement, the Texas-Mexico border region experienced rapid growth which led to towns that lack proper sewage and water treatment systems and has led to public health problems. This study evaluated the use of health education and outreach as a cost-effective means of empowering communities to improve their health. The health promotion program was developed through a specific process. The initial step was to obtain an assessment of existing conditions in the area from multiple perspectives. The second step was to involve the community in the project by letting them choose topics of concern which in this case happened to be safe drinking water. Then the topic was turned into modules that included a short introduction on the issue, why the issue is a concern to health of the families, what can be done at home to reduce the risk of disease or exposure due to the issue, and a prevention section. The material was first developed in English then translated to Spanish. A second translator then reviewed the material and made corrections. Then the community health educators reviewed it and made suggestions to make the material culturally sensitive and reader-friendly. The study delivered the safe drinking water module in two South Texas Colinias, one contained 5,000 families and the other 300 families. Pre and post tests were given prior to the modules. A quantitative data analysis was performed using SPSS statistical software and a 95% confidence intervals for each question on the pre-post test. Significant improvement was seen between the pre and post tests, but it was also noted that there were differences between the educators.

The quality and consistency of the individual educator was critical to the success of the education. When the educators just simply read the information to the families their scores did not improve and on occasionally worsen. Alternatively, when the educator developed a relationship with the family and held a more open discussion of the educational material, the family not only demonstrated improved knowledge with regards to the module, but often provided valuable information on the other health issues in the community. This study also found that families are truly interested in improving their health and only require the appropriate tools and knowledge to make healthy behavioral changes.

Lantagne (2002) shows there is a need for follow-up and education in order for the CWP to be successful. Eun Young Hwang (2003) took Lantagne's (2002) research a step further and reproduced the follow-up and educational need findings but was able to recommend that follow-up and continued education should be aimed at women, literacy rate should be factored in, including the fact that information should be attractive to children because they are often the ones to disperse printed information to their mothers. Roberts's (2003) Cambodia study showed there is higher success rate in populations that already believe water supply is a health issue and are taking action to obtain and maintain a clean drinking water supply. By combining the findings of theses three studies leads one to believe time may be well spent on basic water and disease education even before filters are distributed to help their acceptance in the future.

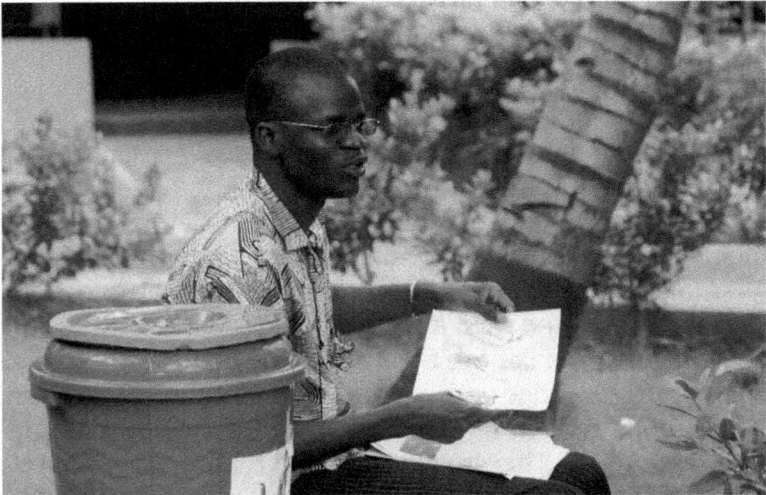

Community participation is needed for sustainable development. The involvement of local people of Benin and Songhai Center in all stages of the project, from planning to construction, and on through to maintenance, encourages a sense of community ownership and responsibility that is needed. It also increases potential for the CWP to be viewed as a local product. In order to achieve the primary objective of improving the health status of the community, there is a need to improve attitudes both with respect to hygiene in home and general health education; this needs to be implemented in conjunction with water supply and environmental sanitation programs. It is not only the technical viability of projects that need to be assessed but also the motivation and commitment of the local people who will be vital to its construction and operation. This is why after the introduction of the filter manufacturing process by PFP and Gonzaga, the program and management is turned over to Songhai for local control and adaptation. A long term goal is to build a health center, which could be a forum for public health education or a place to train local community health educators that the above reviewed studies recommend.

Lantagne's (2002) and Eun Young Hwang (2003) showed higher usage rates of the CWPs were directly related to follow-up visits by a staff member or community leader associated with the sponsoring NGO and that usage rates were significantly lower where no follow-up occurred. This will be a role Songhai assumes as manufacturing is perfected and was a requirement from PFP prior to the start of this project. Public response to risk communication efforts depends not solely on facts and technical information, but also on existing knowledge, values, beliefs, emotions; and whether there is a feeling that the people communicating or managing the risk are trustworthy, credible and similar to the target audience. This is why the partnership with PEP, Gonzaga, and Songhai is so important to the success of the project. Songhai is a leader in the community and a reputable place of business. Another asset Songhai has to offer is its scientific background and personnel who understand the scientific process and the importance of previous research findings as it implements and promotes the CWPs.

Good programming flows from solid understanding of the current situation, a realistic assessment of what is possible and being able to draw on expertise of others. Making sure the promotion works means building on what exists, targeting a small number of risk practices, targeting specific audiences, and identifying the motives for changed behavior. If hygiene promotion is to succeed, it needs to identify and target only those few hygiene practices which are the major sources of risk in any setting. This needs to be observed at the local level and the target communities need to have input. Programs then have to focus their efforts on a small number of messages of proven public health importance to improve message clarity.

Songhai needs to take time and brainstorm a name and logo for CWP. The jar of unity-- maybe a possible symbol for the filter discovered by some of the Gonzaga participants. This was one of King Ghezo's symbols. He was the ninth king of Dahomey (Benin) and ruled 1818 to

1858. The symbol is of a clay jar sieve with holes in it held up by two hands. Ghezo is said to have used the sieve as a metaphor for the kind of unity needed for the country to defeat its enemies and overcome its problems; it takes everyone's hands to pug the holes so the vessel can hold water, and it will take everyone to solve Benin's problems. This could serve as a meaningful cultural symbol for the CWP and the water problem in Benin. However, before implementing any symbol or logo it should first be evaluated by local people in Benin for their reactions. First impression is King Ghezo is a well like part of Benin's history, but history reveals this because he stopped all taxes during his rein. He was able to accomplish this because of his involvement in the lucrative slave trade.

The marketing should be aimed at women and demonstrate the benefits to their lives. It is who women provide most water and food, to support bathing, cooking, household hygiene and cleansing of infants, children, the sick and the elderly (Magee, 2005). As the Cambodia study illustrated, it will save them significant time and money. This is a marketing aspect that should be promoted and could be considered a small step towards the advancement of women. It also has potential to improve the literacy rate by increasing more income for education and decreasing school absentee rates. Families often cannot even afford school supplies to send children to public school. The savings in firewood and time may allow more school attendance.

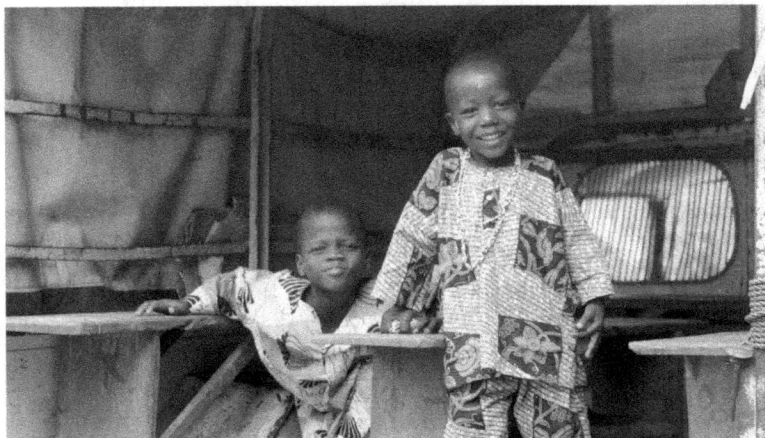

Components an informational flyer should include are: a simple description of the right to and need for clean safe water, the basic cost of a filter and cleaning setup, the benefits of drinking clean water and the savings to the family, information on follow-up visits and a possible health class, where to get the filters, and basic directions of use and proper cleaning. Remember, written information must be designed at a low readability level and should be attractive to children.

The most important aspect the Songhai Center needs to consider at this time is creating a health education position in relation to promoting the CWPs and plan on eventually developing it into a department. It is possible it could also be merged with the marketing side of the CWPs, especially when you take into consideration the studies on health behavior change and social marketing. Before Songhai even starts aggressively marketing the CWP they could be making public service announcements to get the population thinking about the need for change. It would also be valuable to continue to develop the working relationship with the public before the introduction of the CWP.

Reference

Bartram, J., Lewis, K., Lenton, R. & Wright, A. (2005). Focusing on improved water and   sanitation for health. The Lancet, 365, 810-812.

Blantari, J. (2005). An evaluation of the effectiveness of televised road safety messages in   Ghana. International Journal of Injury Control and Safety Promotion, 12, 23-29

Curtis, V., Cairncross, S. & Yonli, R. (2000). Domestic hygiene and diarrhea-pinpointing theproblem. Tropical Medicine and International Health,5,22-32.

Eun Young Hwang, R. (2003). Six-month field monitoring of point-of-use ceramic water filter in   San Francisco Libre, Nicaragua. Massachusetts Institute of Technology.

Gleick, P. (2002). Dirty water: estimated deaths from water-related diseases 2000-2020. Pacific   Institute for Studies in Development, Environment, and Security. 1-12.

Heidecke, C. (2006). Development and evaluation of a regional water poverty index for Benin. International Food Policy Research Institute. 1-38.

Jensen, P., Jayasinghe, G., van der Hoek, W., Cairncross, S., & Dalsgaard, A. (2004). Is there an association between bacteriological drinking water quality and childhood diarrhea in developing countries? Tropical Medicine and International Health, 9, 1210-1215.

Kitawaki, H. (2002). Common problems in water supply and sanitation in developing countries. International Review for Environmental Strategies, 3, 264-273.

Lantagne, D. (2002). Investigation of the Potters for Peace colloidal silver impregnated ceramic filter: Intrinsic effectiveness and field performance in rural Nicaragua. Alethia Environmental. www.alethia.cc

LearnLink. (1999). Reaching remote villages community networking service centers in Benin. Accessed of www.blackboard.gonzaga.edu course 690 summer 2007.

Magee, M. (2005). Healthy waters. Bronxville, NY: Spencer

Nath, K. (2003). Home hygiene and environmental sanitation: a country situation analysis for India. International Journal of Environmental Health Research, 13, S19-S28.

Polit, D., & Beck, C. (2004). Nursing research and method (7th ed.). Philadelphia: Lippincott Williams and Wilkins.

Potters for Peace. (2005). Potters for Peace clay water filter assistance guidelines.   www.potpaz.org. Accessed on August 28, 2007.

Quick, R. (2003). Changing community behavior: experience from three African countries.   International Journal of Environmental Health Research, 13, S115-S121.

Roberts, M. (2003). Ceramic water purifier Cambodia field tests. International Development   Enterprises. www.ide-international.org

Striebig, B. (2005). Azove, Benin, Africa: Site assessment. Prepared for Engineers Without   Borders-USA.

Sumaya, C. (2006). Linking research to health promotion in Texas Colonias. American Journal of Health Studies, 21, 45-53.

Thesenvitz, J. (2000). Risk communication. The UpdateHealth Communication Unit. 2-12.

Thevos, A. (2000). Adaptation of safe water behaviors in Zambia: comparing educational and motivational approaches. Education for Health, 13, 366-376.

United Nations. (2007). Africa and the millennium development goals 2007 upate. UN   Department of Public Information.

United Nations. (2003). Water for People, Water for Life. The United Nations World Water   Development Report. Paris, UNESCO Publishing.

United Nations. (2000). World Population Prospects: The 1999 Revision. New York.

World Bank. (2004). World development indicators, Washington, D.C.: World Bank.

Wright, J., Gundry, S., & Conroy, R., (2004). Household drinking water in developing countries: a systematic review of microbiological contamination between source and point-of-use.   Tropical Medicine and International Health, 9,106-117.

"What I do you cannot do; but what you do, I cannot do. The needs are great, and none of us, including me, ever do great things. But we can all do small things, with great love, and together we can do something wonderful."

— Mother Teresa of Calcutta

Epilogue

Compassion is the first ingredient needed to solve the water crisis. Compassion alone is not enough however. We must also be willing to sacrifice time and risk investing in relationships that take time to develop. As great as the thirst for water is in Africa, the thirst for knowledge is even more acute. By building relationships with partner organization in developing countries, students can change the world. Through these relationships we can begin to understand one another's needs. Only after we educate ourselves through building relationships with our project partners can begin to share a common vision.

The WATER (West African Technology, Education and Reciprocity) study abroad program at Gonzaga was built upon a common vision expressed by Father Nzamujo Godfrey, director of the Songhai Center, and myself. That vision was to bring an effective water treatment technologies technology into homes in Benin.

Manifesting that vision into a sustainable process to make drinking water filters took several years and many, many hours of hard work. It also required building relationships with professors in eight different academic disciplines. This vision had to be shared among faculty and students alike.

Finally in order for the filters to find their way into homes in Benin, students had to educate themselves in basic health care, communication techniques and learn how to make the water filters. Once the WATER team was trained, this team was able to work alongside people at the Songhai Center to train them to make the water filters. An integral part of this training was to develop relationships with our project partners in Benin – everyone had to learn how to work together for a common vision.

As a result of this project, the Songhai Center is currently producing drinking water filters in Benin. In addition, these filters and the water quality is being measured and a health center is under construction. Another WATER tem is preparing to return to Benin during the summer of 2008 to improve water filter performance and conduct basic health care training. Several of the WATER 2007 participants have completed their college degrees. Nearly 50 percent of the WATER team members have entered or plan to enter graduate programs in education, health and engineering.

Students that have worked on the WATER project in Benin during the regular academic year since 2004 have gone onto apply their expertise in the Peace Corps, Jesuit Volunteer Corps, and Engineering Ministries International. Other alumni have gone onto study environmental engineering and water resources at Columbia University, Michigan Technological University, the University of Texas, the University of Washington, and Virginia Technological University. Locally, GU Alumni have founded a chapter of

Engineers Without Borders for professional engineers in the Inland Northwest. Currently Alumni of this program are working in Brazil, the Dominican Republic and Togo, which is also in West Africa.

Finished water filters at the Songhai Center. Photo by Thierry Andre

We all must educate ourselves about the water crisis in order to find the compassion, which will move us to action. The actions we take must be rooted in relationships, sometimes developed in the midst of crisis. Only after we understand one another can we learn to help one another. Only thorough our efforts to help, we will learn how to act humbly based upon our compassion, and find peace both at home and abroad.

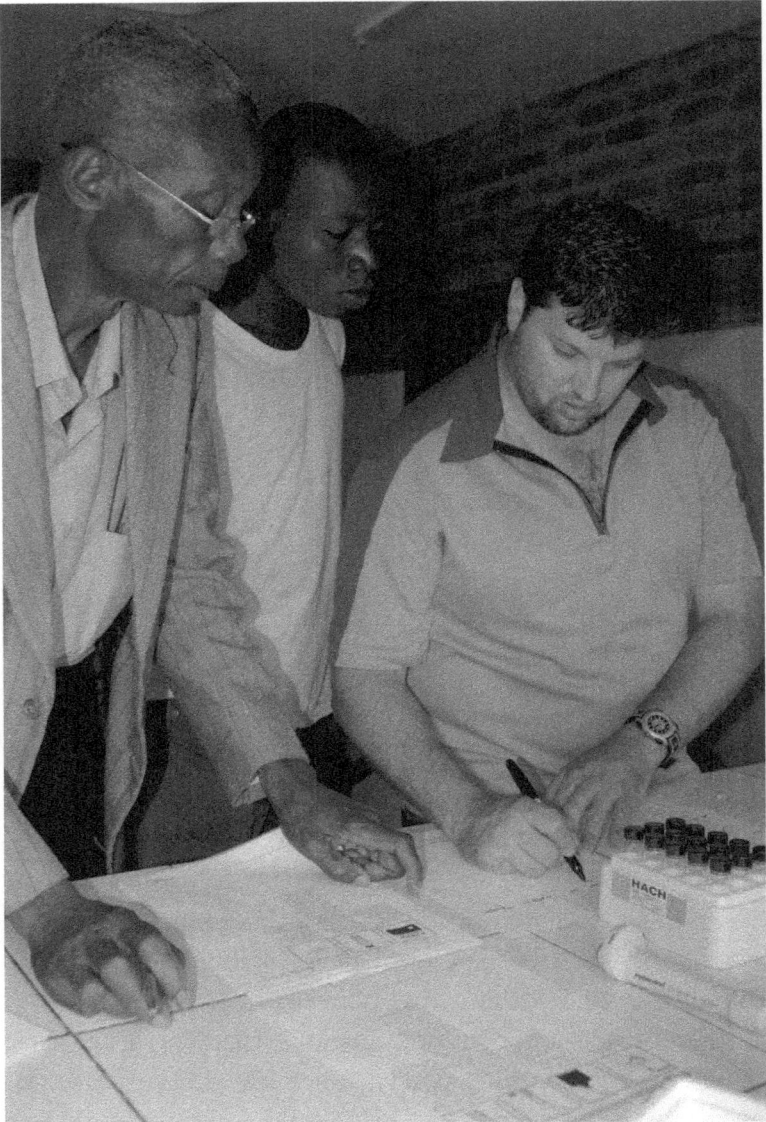

Professor Striebig teaching water analysis techniques at the Songhai Center

About the Author

Dr. Bradley A. Striebig is an associate professor of Civil Engineering at Gonzaga University. He has a Ph.D. in Environmental Engineering from Penn State University, where he worked prior to joining the faculty at Gonzaga. He has worked on various water projects throughout the US and in Benin and Rwanda.

More information about the Songhai Center can be found on their website: **www.songhai.org**

More information about the WATER program can be found online at: **http://web.mac.com/water_dr**

Contributions can be made to the WATER program through the EWB-USA website or by contacting the Gonzaga University School of Engineering and Applied Science at 509-323-3522.

www.ingramcontent.com/pod-product-compliance
Lightning Source LLC
Chambersburg PA
CBHW031520270326
41930CB00006B/455